Dunbing Tang · Kun Zheng · Wenbin Gu

Adaptive Control of Bio-Inspired Manufacturing Systems

Springer

Dunbing Tang
College of Mechanical and Electrical
Engineering
Nanjing University of Aeronautics
and Astronautics
Nanjing, China

Kun Zheng
School of Automotive and Rail Transit
Jiangsu Key Laboratory of Advanced
Numerical Control Technology
Nanjing Institute of Technology
Nanjing, China

Wenbin Gu
College of Mechanical
and Electrical Engineering
Hohai University
Changzhou, China

ISSN 2523-3386   ISSN 2523-3394 (electronic)
Research on Intelligent Manufacturing
ISBN 978-981-15-3447-8   ISBN 978-981-15-3445-4 (eBook)
https://doi.org/10.1007/978-981-15-3445-4

This Springer imprint is published by the registered company Springer Nature Singapore Pte Ltd.
The registered company address is: 152 Beach Road, #21-01/04 Gateway East, Singapore 189721, Singapore

Research on Intelligent Manufacturing

Research on Intelligent Manufacturing (RIM) publishes the latest developments and applications of research in intelligent manufacturing—rapidly, informally and in high quality. It combines theory and practice to analyse related cases in fields including but not limited to:

Intelligent design theory and technologies
Intelligent manufacturing equipment and technologies
Intelligent sensing and control technologies
Intelligent manufacturing systems and services

This book series aims to address hot technological spots and solve challenging problems in the field of intelligent manufacturing. It brings together scientists and engineers working in all related branches from both East and West, under the support of national strategies like Industry 4.0 and Made in China 2025. With its wide coverage in all related branches, such as Industrial Internet of Things (IoT), Cloud Computing, 3D Printing and Virtual Reality Technology, we hope this book series can provide the researchers with a scientific platform to exchange and share the latest findings, ideas, and advances, and to chart the frontiers of intelligent manufacturing.

The series' scope includes monographs, professional books and graduate textbooks, edited volumes, and reference works intended to support education in related areas at the graduate and post-graduate levels.

More information about this series at http://www.springer.com/series/15516

Acknowledgements

This work was supported by the National Natural Science Foundation of China (No. U1637211, 51805244, 51875171), National Key Research and Development Program of China (No. 2018YFE0177000), the Fundamental Research Funds for the Central Universities (No. 2019B21614), National Defense Basic Scientific Research Program of China (No. JCKY201805C003), Jiangsu Province 333 Project, Scientific Research Fund of Nanjing Institute of Technology (No. YKJ201622 and KXJ201606). The authors would like to thank the referees for their helpful comments and suggestions.

Contents

Chapter 1
Bio-Inspired Manufacturing System Model

1.1 Introduction and Synopsis

Nowadays manufacturing enterprises are facing more complex and significant trends of cultural diversification, lifestyle individuality, activity globalization, and environmental consideration. These trends can be summarized as growing complexity and dynamics in manufacturing environments [1], and manufacturing companies are forced to have manufacturing systems that exhibit innovative features to support the agile response to the emergence and changing conditions [2, 3]. In order to meet the new requirements, several manufacturing paradigms have been proposed for the next generation manufacturing, such as agent-based manufacturing system [4–6], fractal manufacturing system [7, 8], holonic manufacturing system [9–11]. These types of architectures are considered to be suited for developing distributed intelligent systems in an open and dynamic environment. However, some problems still remain regarding the complexity of the manufacturing system.

The agent technology and the multi-agent system paradigm have been considered over more than a decade as an important approach for developing and implementing the software components of the intelligent manufacturing system. Multi-agent systems for manufacturing systems appear to provide adequate response to abrupt disturbances on the shop floor. Since there is no central controller, the agents are empowered to manage most of the activities related to their own goals and tasks through intensive inter- and intra-agent communication [12]. The agent participants must know who are in the community and how to communicate with them in advance, therefore, the MAS is likely to be only suited to the well-structured problem and there is no theoretical guarantee that the agent interactions process will ever converge, especially for the large-scale and complex manufacturing system.

© Springer Nature Singapore Pte Ltd. 2020

D. Tang et al., *Adaptive Control of Bio-Inspired Manufacturing Systems*,
Research on Intelligent Manufacturing,
https://doi.org/10.1007/978-981-15-3445-4_1

Holonic manufacturing system (HMS) is another distributed control paradigm that promises to cope with frequent changes and disturbances. HMS is characterized by holarchies of autonomous and cooperative entities, called holons, which represent the entire range of manufacturing entities. The introduction of the holonic paradigm allows a new approach to the manufacturing problem, bringing the advantages of modularity, decentralization, autonomy, and scalability [13]. Fractal manufacturing system (FrMS) has been discussed as a potential candidate for the next generation of manufacturing systems. A FrMS is a new manufacturing concept derived from the fractal factory introduced by Warnecke [14]. It is based on the concept of autonomously cooperating multi-agents referred to as fractals. Most researches of HMS and FrMS still use the agent technology to model holons or fractals, such as [8, 11, 15], and these researches have been performed under the similar banner of "agent-based manufacturing". Therefore, HMS and FrMS cannot avoid the problems of the multi-agent system, and especially how to achieve global optimization in decentralized manufacturing systems is still an open question.

Although multi-agent manufacturing system, HMS and FrMS have different underlying operational principles, there is a common character that they advance an organization of autonomous modules that try to be capable of self-organizing and adaptive behaviors to carry out the necessary functions in the changing environment. Based on the common sense that biological organisms naturally can be capable of adapting to environmental changes, researchers have begun to pay attention to the Bionic Manufacturing System (BMS) [1, 16]. One typical type of BMS research is focused on imitating the biological evolution through generation (DNA-type) using evolutionary computation algorithms. Another kind of arresting BMS research is to make use of pheromone-like techniques (such as pheromones deposited by ants) to build the biologically inspired manufacturing system [17]. It is no doubt that current BMS researches have gained some meaningful results, while the achievements leave an important question open: can we mimic the advanced controlling, regulating, and modulating mechanism of biological organism for bio-inspired manufacturing systems?

In a biological body, the neuroendocrine system is commonly regarded as one of the major physiological systems and has some special modulating mechanism for better control adaptability and stability [14, 18, 19]. The neuroendocrine system exhibits neuro-control and hormone regulation behaviors. By referencing the neuroendocrine mechanism, a new Bio-Inspired Manufacturing System (BIMS) is proposed. The aim of this research is to improve the intelligence, controllability, and adaptation of the manufacturing system by utilizing the mechanism of neuro-control and hormone regulation, thus consolidating and deepening the BMS theory and fundamentals.

1.2   The Biological Background of BIMS

Nervous system, endocrine system, and immune system are important physiological systems used by biological organisms to adjust bodies to changes in the internal and external environment. Take the human body system as an example, three sub-systems and their mutual adjustment relations are introduced in the following sub-sections.

1.2.1   Nervous System

The Nervous system is the dominant system of the human body. It is a vast complex network that controls every aspect of human life and behavior. This system is inter-woven all over the body, receiving, decoding, and taking actions to the information obtained from the outside and itself, so as to maintain the homeostasis and balance of the internal environment of the body. The nervous system can be divided into two categories: peripheral nervous system (PNS) and central nervous system (CNS), as shown in Fig. 1.1.

Due to the dominance of the nervous system, the other organs are subordinate. The nervous system innervates and regulates the activities of each organ through the nervous reflex. The human body is full of various nerve cells that can receive stimulation and transmit excitement, which is the basic unit of the nervous system. Environmental changes outside the body can stimulate these nerve cells to produce stimulus signals, known as nerve impulses. Nerve impulses travel through neurons to the nerve center. After analysis, the nerve center produces nerve signals that travel through neurons to the effectors and act on the corresponding organs.

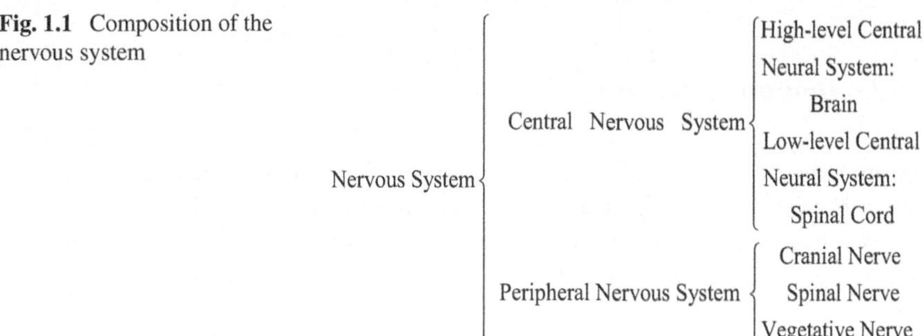

Fig. 1.1 Composition of the nervous system

1.2.2 Endocrine System

The endocrine system is a chemical messenger system and can secrete some chemical substances into the body, which affect the functions of certain tissues. The chemical substances are called hormones; the tissues that secrete chemical substances are called endocrine glands; the tissues inside the body that can receive and be affected by hormones are called target organs; and hormones are transmitted through the blood circulation system. Hormones play an important role in the normal functioning of the body, among which growth hormones, sex hormones, and thyroid hormones are the common hormones. For instance, thyroid hormones can rapidly absorb sugar, promote the synthesis of proteins in the body and regulate the excitability of the nervous system.

The endocrine system secretes hormones including endocrine, neuroendocrine, paracrine, and autocrine. Endocrine is a process of hormone secretion by glands, afterward hormones enter the blood circulation system and are acted on target organs. The process of neuroendocrine secretes neurohormones which stimulate endocrine glands to perform corresponding endocrine activities. Paracrine is a process of hormone secretion by glands that stimulates target organs nearby. Autocrine is a process of hormone secretion by endocrine glands that spreads around glands and feeds back into themselves for secreting hormones.

The endocrine system consists of various types of endocrine glands. Endocrine glands consist of receptors and effectors. Different endocrine glands have their own receptors, which can sense the stimulation of certain hormones in the body-environment then promote or inhibit the release of hormones through effectors. The endocrine system utilizes the interactions between endocrine glands and adapts to changes in internal and external environments through the regulation of hormone environment.

1.2.3 Immune System

The immune system is the main system performing defense functions inside the human body that can protect antigenic foreign bodies (such as bacteria, viruses, tumors, etc.), and produces specific physiological responses. The system can recognize antigenic foreign bodies and exclude or eliminate them to protect the body's health.

The Immune system is a complex adaptive system composed of immune tissues and organs, immune cells and molecules, which is shown in Fig. 1.2.

Immune organs are places where immune cells get activated and mature, including the thymus and bone marrow. Peripheral immune organs are the places where T and B cells settle and develop immune responses, including lymph nodes, spleen, and mucosa-associated lymphoid tissue. The main functions of the immune system are as follows:

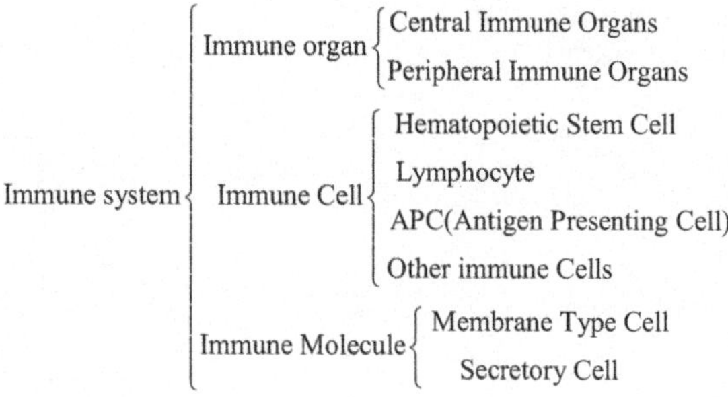

Fig. 1.2 Components of the immune system

(1) Immune defense: a function of the body is to recognize and eliminate antigenic foreign bodies and maintain health. In the process of immune defense, if the immune response is too strong or lasts too long, it will cause damage to its own tissue. If the immune response is too weak or immunodeficient, it will make it difficult for the immune system of the body to remove the antigenic foreign body.

(2) Immune surveillance: it refers to a function of the body to recognize and exclude abnormal mutant cells. During the metabolism of cells, some cells may be mutated or distorted due to various factors. If these mutant cells cannot be excluded in time, there is a possibility of mutant cells becoming a tumor.

(3) Immune self-stabilization: it refers to the body's recognition and exclusion of damaged or thrown out of its own cells through immune regulation to maintain the stability of body functions

In addition to the above three functions, the immune system also includes immune learning, immune recognition, immune memory, and other functions.

1.2.4 Neuroendocrine-Immune System

In the human body, the nervous system, endocrine system, and immune system exist bidirectional transmissions and acting mechanisms among each other. Their interactions are mainly realized through chemical signal molecules and receptors in the nervous system, endocrine system, and immune system. Here, chemical signal molecules refer to neurotransmitters, hormones, and immune factors. The three systems work together, regulate each other, maintain the internal balance of the human body, and make the body work in a normal state, and their mutual relations are shown in Fig. 1.3.

Fig. 1.3 Interactions of
three sub-systems

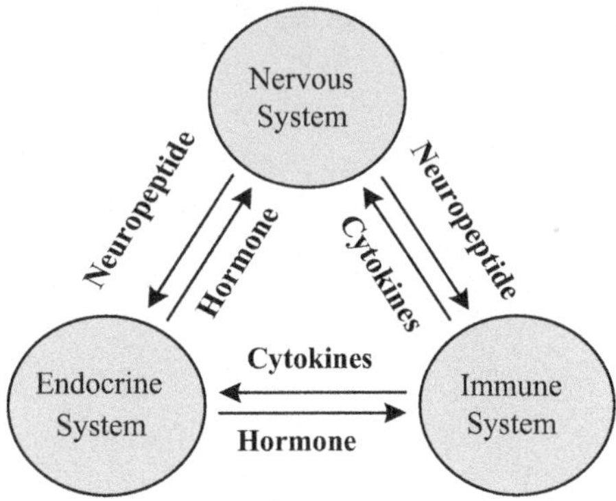

1.2.4.1 The Mutual Regulation of Nervous System and Endocrine System

The interactions between the nervous system and the endocrine system are the main research topic of the neuroendocrine system and its simple model shown in Fig. 1.4.

The function and structure of the nervous system and endocrine system are interrelated through interactions between the hypothalamus and pituitary. As the center of endocrine regulation, the hypothalamus receives nerve stimulation from the central nervous system, secretes nerve hormones, directly acts on the pituitary, and regulates endocrine functions of pituitary. As an endocrine gland, pituitary can secrete various hormones, act on each gland in the body, and regulate hormone secretion of these glands. Many functions of the "hypothalamic-pituitary" system are spread across other systems and regulate most metabolic activities in the body.

In some cases, the nervous system can directly stimulate and regulate certain endocrine glands. For example, when the human body encounters an emergency situation, sympathetic nerve activities increase abruptly and stimulate endocrine glands to secrete adrenaline to promote the body accelerating heart rate and increasing blood pressure in response to external emergencies. The endocrine system can regulate the function of the nervous system through hormones. Specifically, endocrine hormones can regulate the conduction efficiency of nerve impulses between nerve cells, change the sensitivity and response of nerve cells to external stimuli, and thereby affect the functions and activities of the nervous system. For instance, thyroid hormones play an important role in stimulating and regulating neural development, and inadequate levels of hormones can affect the development of the nervous system. To sum up, the effect of the mutual regulation of the nervous system and the endocrine system on the body system is reflected in the following aspects

(1) Internal environmental balance: through the regulation of neuroendocrine, the internal biochemical environment of the body can be maintained in a relatively stable state and dynamic balance.

(2) Biorhythm: ceaseless interactions between the nervous system and the endocrine system keep nervous activities and fluctuation of hormone levels, which affects

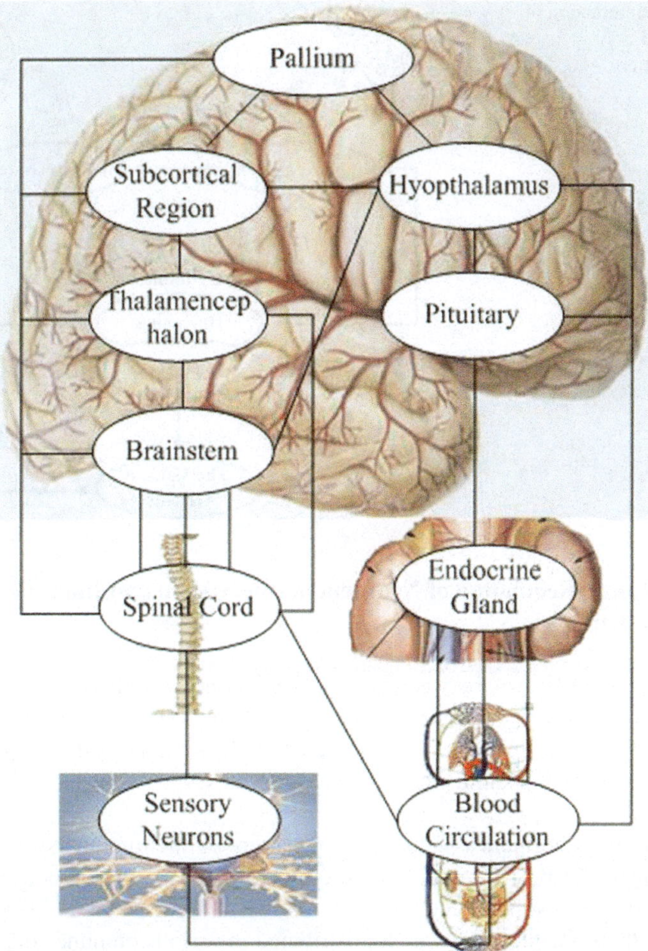

Fig. 1.4 General model of the neuroendocrine system [23]

activities and functions of the human body, thus forming the human body's biorhythm.

(3) Stress response: when a body is exposed to harmful stimuli, on one hand, the nervous system responds rapidly through neural control and adjusts the body to avoid danger; on the other hand, the neuroendocrine system stimulates the secretion of adrenaline and improves stress response to an adverse environment.

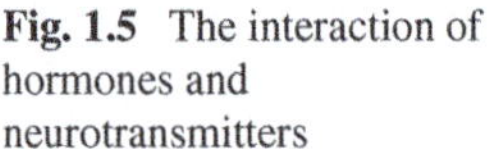

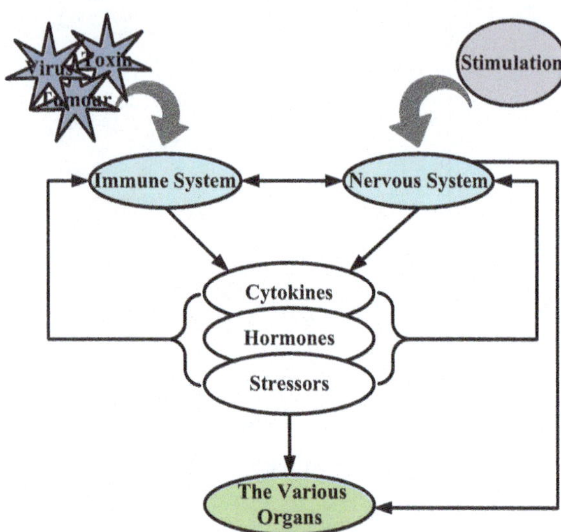

Fig. 1.5 The interaction of hormones and neurotransmitters

1.2.4.2   Mutual Regulation of Neuroendocrine System and Immune System

The interaction mechanism between the neuroendocrine system and the immune system is shown in Fig. 1.5. Through the interaction of hormones and neurotransmitters, the neuroendocrine system can have an impact on the immune system. This mechanism can be reflected in some physiological processes, such as lactation, stress, and pregnancy.

Most effects of hormones and neurotransmitters on the immune system present a single enhancement or inhibition. And only two hormones (Adrenocorticotropic Hormone (ACTH) and beta-endorphins) have different effects on different immune cells, while their receptors are widely distributed in various immune cells. So, they have both inhibition and enhancement effects on the immune system. The immune system can affect the neuroendocrine system through a variety of pathways. Table 1.1 and Table 1.2, respectively, show the effects of neuroendocrine hormones and classic neurotransmitters on immune function. Studies have shown that immune cells can not only produce and release endocrine hormones to affect the neuroendocrine system, but also generate a variety of cytokines that act on the whole-body organs and the neuroendocrine system.

The immune system is generally considered to be a part of the neuroendocrine system. When the body is stimulated by viruses or toxins, immune cells secrete and release neuroendocrine peptide hormones, known as an immune reactive hormone. Immune cells activated by external influences release a variety of cytokines, which can not only regulate the activity of immune cells themselves , but also influence

Table 1.1 Impacts of hormones on the immune system

Hormone	Basic function	Specific effects
Glucocorticoid	Inhibition	Antibodies, production of cytokines, activity of natural killer (NK) cells
Thyroid hormones	Potentiation	Proliferation of thymus cells, lymphocytes, and spleen cells
Vasopressin	Potentiation	T cell proliferation
ACTH	Inhibition/potentiation	Antibodies, production of cytokines, NK and macrophage activity
HGH	Potentiation	Antibody synthesis macrophage activity
Androgenic hormone	Inhibition	Lymphocyte transformation
Estrogen	Potentiation	Lymphocyte transformation
Corticotropin releasing hormone (CRH)	Potentiation	Production of cytokines
Prolactin	Potentiation	Antibody synthesis, thymic tissue hyperplasia, NK and macrophage activity
ß-endorphin	Inhibition/potentiation	Antibody production, T cell, and macrophage activity

Table 1.2 The impacts of neurotransmitters on the immune system

Neurotransmitter	Basic function	Specific effects
Norepinephrine	Potentiation	Enhance humoral immunity and inhibit proliferation of T cells
Acetylcholine	Inhibition	Enhancing humoral immune response and inhibiting proliferation of T cells
5-serotonin	Inhibition	Mediators affecting sympathetic activity of lymphocytes

the neuroendocrine system and regulate the whole-body system. Although cytokines can regulate the neuroendocrine system through different pathways, sometimes these effects are quite similar.

1.3   Bio-Inspired Manufacturing System (BIMS)

The structure and workings of natural life exhibit autonomous and spontaneous behavior [20, 21]. For instance, without the explicit order from the brain, the heart organ seemingly acts on its own while coordinating its actions and maintaining harmony. In the biology area, it is already known that the neuroendocrine system is quite critical in initiating life sustaining adaptive reactions to internal (disease) and external (environmental) stressors. Based on the general concept of Bionic Manufacturing System (BMS), hereby we consider the manufacturing system as a living organism. To be different from other BMS researches, a bio-inspired manufacturing system is presented based on the general principles of the neuroendocrine system. In this section, we will first briefly introduce the general principles of neuroendocrine system, and then propose a novel architecture of the Bio-Inspired Manufacturing System (BIMS) [22].

A biological viewpoint has close parallels in manufacturing. Inspired by the introduced principles of the neuroendocrine system, a new architecture of the bionic manufacturing system is proposed in Fig. 1.6.

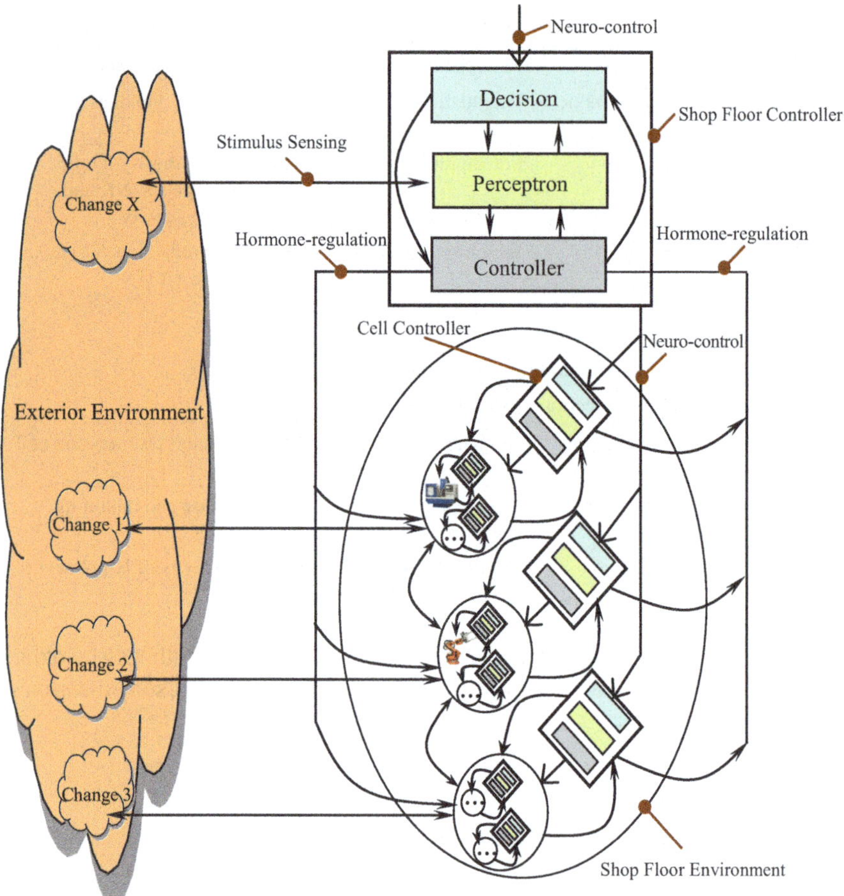

Fig. 1.6 Architecture of BIMS

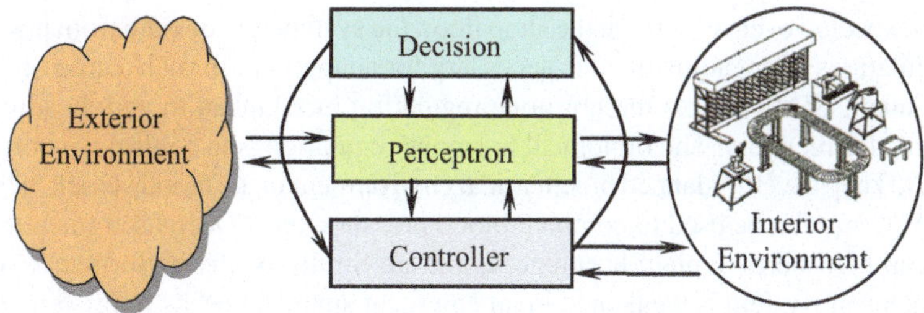

Fig. 1.7   Internal structure of bio-inspired manufacturing cell (BIMC)

The architecture is built upon a set of autonomous and cooperative units whose operations follow the principles of neuro-control and hormone regulation. The composing unit of BIMS is defined as Bio-Inspired Manufacturing Cell (BIMC) which is shown in Fig. 1.7. Each BIMC contains the functional components of controller, decision-maker, and perceptron. Like biological sensory neurons, the perceptron of BIMC can apperceive the exterior and interior environmental stimulus, which triggers the decision of an agile reaction. After a decision is taken, the controller will give a command to the manufacturing device and resource to execute the decision.

A BIMC is defined as an autonomous entity that can perform some tasks and achieve a goal autonomously, and can regulate itself when facing non-predeterministic changes in the manufacturing environments. The design of a basic BIMC incorporates a set of pertinent attributes that can fully represent any level in the architecture hierarchy. In other words, a BIMC can represent an entire manufacturing shop at the highest level or a physical machine at the bottom-level. The architecture, therefore, exhibits an explicit characteristic of recursivity. To be different from HMS and FrMS which are designed to share similar structural features at different levels, BIMS is controlled by quasi-neuroendocrine principles for meaningful integration and agile adaptation from level to level. Acting as the central nervous system (CNS), the shop floor controller conducts the role of the supervisor to give commands to the cell controllers of low-level BIMCs. Under the neuro-control mechanism, the cell controllers can be compared to the peripheral nervous system (PNS). PNS conducts impulses from the CNS to effectors (muscles and glands). Similarly, the cell controllers execute the commands from the shop floor controller to invoke the operation of physical resources or devices. The shop floor environment can be compared to a blood circulation system, and the task schedule and resource (equipment) utilization are regarded as two types of hormones of BIMS.

Based on the principles of the neuroendocrine system, BIMS conducts neuro-control in a normal state, and applies hormone regulation in an emergent state. In a normal state, BIMCs at different levels are organized in a hierarchical structure, and the shop floor controller (similar to CNS) elaborates and sends schedule plans (similar to signals) to the cell controllers of BIMCs. BIMCs at different levels follow the received command and conduct their own stationary operations (just like the heart pumping blood with a normal rhythm).

If unexpected events occur on the shop floor, the system will deviate from planned, and regulationsystem adjustment is necessary for adaptation. In such a case of emergent context, BIMS adopts the hormone regulation mechanism to agilely adjust its behaviors for recovery. The biological hormone regulation is to control the hormone release to keep a well-balanced biochemical environment in the blood vessel. "Higher (lower)" terms can be used to compare blood pressure, and "Decreased (increased)" terms can be used to explain hormone secretion. Similarly, the performance of the manufacturing system is weakened in an emergent state and needs improvement. In other words, the resource shall be re-organized, and/or the original task schedule plan shall be adjusted for re-schedule. Taking the task schedule and resource (equipment) utilization as hormones of BIMS, a quasi-hormone regulation model of BIMS is set up and illustrated in the following section. The basic principle is to modulate the resource utilization to obtain an optimized task re-schedule. The BIMC which detects a disturbance initiates the recovery locally from the emergence by sensing the concentration oscillation of hormone (such as machine failure will negatively affect the resource utilization). Meanwhile, the initiated BIMC interacts with other ones for cooperation, as its disturbance surely affects the original task schedule. To handle the reaction to disturbance and maintain the system stable and to handle the reaction to disturbance, with the supervision of shop floor controller, related BIMCs will regulate their operation burden through cell controllers in order to achieve an alternative schedule plan to ensure on-time product delivery while keeping low work-in-process inventory. In such a context, the cell controllers conduct the role of the pituitary, and the operational devices of BIMCs are similar to the glands. The shop floor controller still acts as the central nervous system. It doesn't directly deal with the disturbance, but receives the feedback of the interactions among the involved BIMCs, and then continues elaborating the task allocation for recovery.

Compared with FrMS and HMS, BIMS has an advantage on agile regulation and coordination in volatile environments. In FrMS or HMS, the coordination is normally conducted through message communication. That's to say, an initiated entity (such as agent or holon) actively sends a message to another one, the receiving side may further propagate the message to other entities. If the system scale is large and complex, the load of message communication would be quite heavy and system agility cannot be guaranteed. In BIMS, the neuroendocrine principles can enable BIMCs to sense the resource-related and task-related hormone concurrently at any moment, and the hormone regulation mechanism can enable synchronous interactions of involved BIMCs for an agile reaction.

1.4 Control Model of BIMS

One objective of BIMS is to develop an adaptive control approach that enables the agility and reaction to unexpected changes and guaranteeing a global optimization. As stated in the section above, the hormone regulation mechanism has been adopted

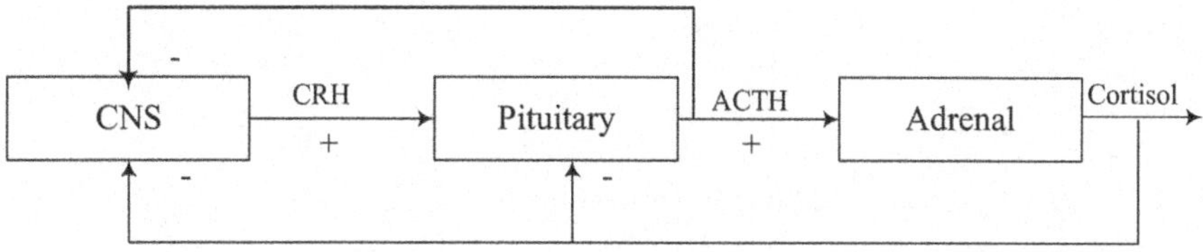

Fig. 1.8 Negative feedback control model of hormone regulation

by BIMS to react to disturbances in an emergent state. In this section, the control model of BIMS is illustrated with a focus on hormone regulation.

1.4.1 Biologic Hormone Regulation Mechanism

The endocrine system is made up of a series of ductless glands that produce chemical messages called hormones. A number of glands that signal each other in sequence are usually referred to as an axis, for example, the hypothalamic-pituitary-adrenal axis. In this sub-section, a feedback-controlled ensemble model of the stress-responsive hypothalamic-pituitary-adrenal axis [19] is used to explain the hormone regulation mechanism and related control model. This neuroendocrine ensemble exhibits prominent time-dependent dynamics reflected in the vividly pulsatile and 24-h rhythmic (circadian) output. Episodic secretion is driven by hypothalamic neuronal pacemakers of CNS, which secrete the pituitary signaling peptides CRH (ACTH-releasing hormone). These agonists singly and synergistically stimulate ACTH (adrenocorticotropic hormone) synthesis and secretion (feed-forward), which, in turn, promotes the dose-responsive biosynthesis of cortisol. Cortisol feedback to inhibit CRH and ACTH production via concentration-dependent and rapid rate-sensitive mechanisms. In other words, when the level of cortisol concentration is too high, it feeds back on the hypothalamus and pituitary to shut down the secretion of CRH and ACTH, which inhibits cortisol secretion in accordance and leads to a biochemical balance. A typical control model of hormone regulation is shown in Fig. 1.8, which exhibits the characteristic of negative feedback.

1.4.2 Hormone Regulation Model of BIMS

Inspired by the biologic hormone regulation mechanism, a control model of BIMS for coordination is proposed in Fig. 1.9. The notations of this model are

$\overline{T}$	global task of the original plan
$\overline{R}$	global resource capacity of the original plan
$\overline{T}_i$	task plan to BIMC i in the original plan
$\overline{R}_i$	resource allocation to BIMC i in the original plan

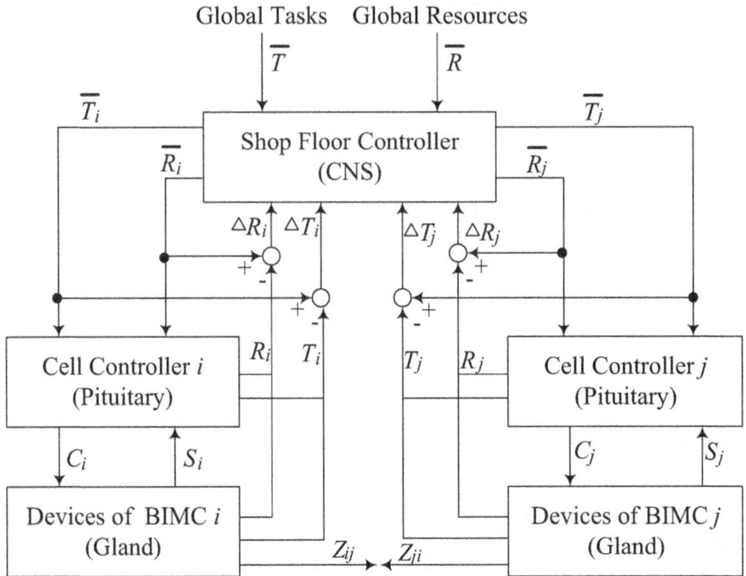

Fig. 1.9 Control model of BIMS

T_i task finished by BIMC i

R_i resource capacity consumed by BIMC i

ΔT_i task deviation of BIMC i

ΔR_i resource capacity deviation of BIMC i

C_i control order to BIMC i from cell controller i

S_i state of BIMC i

$Z_{ij}(Z_{ji})$ machine skill similarity (0–1) between BIMC i and BIMC j

Mimicking the hormone regulation principle, the shop floor controller is regarded as the CNS, the cell controller as the pituitary, and the devices of BIMC as the gland (such as adrenal). Taking the resource utilization and task schedule as two types of hormone, the level of hormone concentration of BIMC i can be indicated as ΔR_i and ΔT_i. During the schedule plan, the information on resource utilization contains machines' type, capability, and a number of machine units; and the task schedule refers to the operations assigned with the required amount of time for processing on the specified type of machines. Therefore, resource utilization and task scheduling are interdependent on each other. Suppose that the resource hormone is controlled by the task hormone. If ΔT_i is not equal to zero, it means unexpected events occur on the shop floor, and the hormone regulation will be triggered for system equilibrium through resource regulation. The resource regulation is to re-organize the resource allocation for re-scheduling, and regulated resource capacity $\delta \overline{R}_i$ (like the secreted hormone for biological regulation) will be assigned to BIMC i.

The basic condition of resource regulation can be described as

$$\sum_{1}^{n} \delta \overline{R}_i + \sum_{1}^{n} \Delta R_i \geq 0 \tag{1.1}$$

$$\delta \overline{R} = \left(\sum_{1}^{n} \delta \overline{R}_i \right) \geq 0 \tag{1.2}$$

$$\Delta R_i = \overline{R}_i - R_i \tag{1.3}$$

$$\sum_{1}^{n} \overline{R}_i + \delta \overline{R} = R \tag{1.4}$$

where $\delta \overline{R}$ is the summation of regulated resource capacity, R is the total consumed resource capacity after regulation.

In the biological hormone regulation mechanism, normally the release of one type of hormone is up-(down-) regulated by another type of hormone. Inspired by such a mechanism, it is considered that the resource capacity regulation is controlled by the task deviation, and the relationship between them can be described as

$$\delta \overline{R}_i = K_R |\Delta T_i| \mathrm{sign}(\Delta T_i) \tag{1.5}$$

where $\delta \overline{R}_i$ is the regulated resource capacity of BIMC i, $\Delta T_i = (\overline{T}_i - T_i)$ is the task deviation of BIMC i, $K_R > 0$ is the regulation coefficient, and

$$\mathrm{sign}(\Delta T_i) = \begin{cases} +1 & \text{if } (\Delta T_i > 0) \\ 0 & \text{if } (\Delta T_i = 0) \\ -1 & \text{if } (\Delta T_i < 0) \end{cases}$$

To quantitatively describe the regulated resources and task deviation, $\delta \overline{R}_i$ is expressed with the number of type x resource units ($x \in X$, and X is the number of type x resource available), ΔT_i is expressed with the required amount of time originally planned for processing on type y resource, and K_R is determined based on the involved resource capability and expressed with unite time resource.

Based on the hormone regulation mechanism, resource regulation is trying to minimize deviations from the initial plan. $\Delta T_i > 0$ means that BIMC i executes earlier operations and its tasks are finished ahead of schedule, or the existing order/job is canceled. In this context, the cell controller i will conduct resource capacity repression to keep step with other BIMCs. $\Delta T_i = 0$ means that the original task plan can be ensured and there is no need to regulate the resource capacity. $\Delta T_i < 0$ means that an unexpected disturbance (e.g., a machine failure, rush orders) is detected, and BIMC i cannot meet the due date. There are many sources of uncertainty in real-world manufacturing system, which may trigger disturbance events on the shop floor. Generally speaking, there are two types of disturbances, namely, resource-related disturbance, and source-related disturbance [13]. Resource-related disturbance refers to the disturbance caused by unreliability coming from resources (machines) in the shop, including machine breakdown and machine recovery. Source-related disturbance refers to the disturbance caused by the changes in production orders, including new order/job arrival. In such context, the cell controller will act as a resource activator to release affordable resources. One way of resource activation is to utilize the redundant resources of BIMC i. Another way is to ask resource help from other BIMCs with similar machine skills.

After the execution of local resource regulation of BIMCs performed in a distributed manner, it is necessary to synchronize and optimize a global schedule. The synchronization is conducted by the shop floor controller under the global resource constraint. By applying hormone regulation principle to deal with malfunctioning events, this approach reduces the loading of the shop floor controller by empowering each BIMC controller to determine the job re-schedule of its respective cell. On the other hand, like the CNS, the shop floor controller still takes charge of all the tasks related to the global shop resource conditions for minimizing the task finishing time, which can be described as

$$\left. \begin{array}{c} T(T_1, T_2, \ldots, T_n) \rightarrow \min \\ \text{s.t.} \quad \sum_1^n R_i \leq \overline{R} \end{array} \right\} \tag{1.6}$$

After the recovery from the disturbance, the cell controllers end the resource regulation, and the system is evolving to a new control state (often returning to the original one). The shop floor controller returns to its neuro-control function, and continues elaborating and proposing the allocation of work orders to the cell controllers of BIMCs.

1.5 Conclusion

In the current manufacturing paradigms (including multi-agent manufacturing system, HMS, and FrMS), most of the popular coordination mechanisms are based on the agent or multi-agent coordination and control, and these approaches are kinds of direct negotiation mechanisms in which participants must know who is in the community and then ask others explicitly for cooperation. Such communication overhead decreases the agent's decision-making response to disturbances in the system, and makes agents spend more time processing messages than doing the actual tasks; therefore, the system's performance will decrease.

Biological organisms are quite capable of adapting to environmental changes and stimulus, and it has been recognized that the neuroendocrine system plays a quite important role to control and modulate the adaptive behaviors of biological organisms using neuro-control and hormone regulation principles. This chapter, therefore, proposes a novel concept of BIMS. Our long-term goal is to develop a neuroendocrine-inspired methodology to attain high-level adaptability of the manufacturing system. As the first step toward this goal, in this chapter,, we introduce a novel BIMS architecture that imitates the principles of neuro-control and hormone regulation to agilely deal with the malfunctioning disturbances on the shop floor. The BIMS architecture is composed of BIMCs at different levels, and cell controllers manage most of the information flow that occurs in the respective BIMCs. As the topic of BIMS is very new, the following chapters will allow us to focus our attention and achieve better understandings about BIMS without being buried in a complicated manufacturing problem itself.

To end this chapter, we discuss the limitations of this research and the extension of future research. First, the inherent limitation of a simple case study should be noted. Given the relatively simple case study, the external generalizability of the findings is limited. Real manufacturing systems can be really complex systems. Future research can address this limitation by examining BIMS in real manufacturing companies. Second, in the pilot implementation of BIMS, it is mainly focused on malfunctioning events with machining resources on the shop floor rather than attempting a full control of manufacturing activities. However, job scheduling shall consider the dynamic events related to human resources, buffer equipment, etc. Especially, work-in-process (WIP) may occur and additional storage costs will need to be considered. Third, a methodology is needed to measure the controller performance of the proposed BIMS architecture. The evaluation results can be used as a reference to the further development of the BIMS control model when there are certain new functions included in the future.

References

1. Ueda, K., Vaario, J., & Ohkura, K. (1997). Modeling of biological manufacturing systems for dynamic reconfiguration. *Annals of the CIRP, 46,* 343–346.
2. Wiendahl, H. P., & Scholtissek, P. (1994). Management and control of complexity in manufacturing. *Annals of the CIRP, 43,* 533–540.
3. Leitao, P. (2008). A bio-inspired solution for manufacturing control systems. In A. Azevedo (Ed.), *Innovation in manufacturing* (pp. 303–314) Boston: Springer.
4. Shen, W., & Norrie, D. H. (1999). Agent-based systems for intelligent manufacturing: A state-of-the-art survey. *Knowledge and Information Systems, 1*(2), 129–156.
5. Brennan, R. W., Fletcher, M., & Norrie, D. H. (2002). An agent-based approach to reconfiguration of real-time distributed control systems. *IEEE Transactions on Robotics and Automation, 18*(4), 444–451.
6. Wang, D. S., Nagalingam, S. V., & Lin, G. C. I. (2007). Development of an agent-based Virtual CIM architecture for small to medium manufacturers. *Robotics and Computer Integrated Manufacturing, 23*(1), 1–16.
7. Ryu, K., & Jung, M. (2003). Agent-based fractal architecture and modeling for developing distributed manufacturing systems. *International Journal of Production Research, 41*(17), 4233–4255.

8. Ryu, K., & Jung, M. (2003). Modeling and specifications of dynamic agents in fractal manufacturing systems. *Computers in Industry, 52*(2), 161–182.

9. Brussel, H. Van, Wyns, J., Valckenaers, P., Bongaerts, L., & Peeters, P. (1998). Reference architecture for holonic manufacturing systems: PROSA. *Computers in Industry, 37*(3), 255–274.

10. Leitao, P., & Restivo, F. (2006). ADACOR: A holonic architecture for agile and adaptive manufacturing control. *Computers in Industry, 57,* 121–130.

11. Colombo, A. W., Schoop, R., & Neubert, R. (2006). An agent-based intelligent control platform for industrial holonic manufacturing systems. *IEEE Transactions on Industrial Electronics, 53*(1), 322–337.

12. Nahm, Y.-E., & Ishikawa, H. (2005). A hybrid multi-agent system architecture for enterprise integration using computer networks. *Robotics and Computer-Integrated Manufacturing, 21,* 217–234.

13. Xiang, W., & Lee, H. P. (2008). Ant colony intelligence in multi-agent dynamic manufacturing scheduling. *Engineering Applications of Artificial Intelligence, 21,* 73–85.

14. Warnecke, H. J. (1993). *The fractal company: A revolution in corporate culture.* Berlin: Springer.

15. Deen, S. M. (2003). *Agent-based manufacturing: Advances in the holonic approach.* Berlin: Springer.

16. Okino, N. (1994). Bionic manufacturing system. *Journal of Manufacturing Systems, 23*(1), 175–187.

17. Wang, L., Tang, D. B., Gu, W. B., et al. (2012). Pheromone-based coordination for manufacturing system control. *Journal of Intelligent Manufacturing, 23*(3), 747–757.

18. Farhy, L. S. (2004). Modeling of oscillations of endocrine networks with feedback. *Methods Enzymology, 384,* 54–81.

19. Keenan, D. M., Licinio, J., & Veldhuis, J. D. (2001). A feedback-controlled ensemble model of the stress-responsive hypothalamo-pituitaryadrenal axis. *PNAS, 98*(7), 4028–4033.

20. SureshKumar, N., & Sridharan, R. (2009). Simulation modeling and analysis of part and tool flow control decisions in a flexible manufacturing system. *Robotics and Computer Integrated Manufacturing, 25,* 829–838.

21. Tharumarajah, A., Wells, A. J., & Nemes, L. (1996). Comparison of the bionic, fractal and holonic manufacturing system concepts. *International Journal of Computer Integrated Manufacturing, 9*(3), 217–226.

22. Tang, D., Gu, W., et al. (2011). A neuroendocrine-inspired approach for adaptive manufacturing system control. *International Journal of Production Research, 49*(5), 1255–1268.

23. Tu, X. Y., Wang, Z., & Guo, Y. W. (2005). *Large systems cybernetics.* Beijing: Press of Beijing University of Posts and Telecommunications.

Chapter 2
Hormone Regulation Based Algorithms for Production Scheduling Optimization

2.1 Introduction and Synopsis

Scheduling involves the allocation of resources over a period of time to perform a collection of tasks. It is a decision-making process that plays an important role in most manufacturing and service industries. An effective schedule enables the industry to utilize its resources effectively and attain the strategic objectives as reflected in its production plan. The job-shop scheduling problem (JSP), a typical production scheduling problem, is one of the existing combinatorial optimization problems and it has been well known as an NP-hard problem [1]. The classical JSP can be described as follows: There are n jobs and m machines. Each job consists of m operations, and every operation, that the processing time is already known, should be processed on different machines. At the same time, each machine can only process one job, which cannot be interrupted. Finally, the main objective of JSP is to find an optimization schedule such that makespan is minimized.

JSP can be applied to the manufacture processing and effects the production time and the cost of production for a plant [2]. During the past few decades, JSP has attracted wide research attention, because it is non-deterministic polynomial-time hard (NP-hard), and it is difficult to find an exact solution in a reasonable computation time with the simple optimization algorithm. Many approximate methods, especially the evolution algorithms, have been developed to solve the problem. Reeves [3] proposed a heuristic algorithm to find an optimal solution with a complex model. One of the heuristic algorithms called meta-heuristic algorithm with random number search techniques is used in a very wide range of practical problems. Early generic meta-heuristic algorithms include genetic algorithm and tabu search method [4]. By far, a lot of other algorithms, such as the ant colony algorithm, particle swarm algorithm, differential evolution algorithm, firefly algorithm, etc., all of which are the imitation methods of nature or biological circles, being applied in the job-shop scheduling problem (JSP) and achieved good results. In this chapter, we focus on exploiting two improved evolution algorithms (PSO and GA), which are inspired by

© Springer Nature Singapore Pte Ltd. 2020
D. Tang et al., *Adaptive Control of Bio-Inspired Manufacturing Systems*,
Research on Intelligent Manufacturing,
https://doi.org/10.1007/978-981-15-3445-4_2

the regular mechanism of the neuroendocrine system, to achieve a better solution for JSP.

Genetic algorithm (GA) is one of the popular meta-heuristics that is based on the genetic evolution mechanism of biology. One of GAs' main characteristics is to directly operate on the problem structure without derivation and function continuity limitation. GAs also have the inherent implicit parallelism and global searching ability and can adjust search directions automatically and self-adaptively. The original GA was used to JSP by Hsiang and Cherng [5], who formed a preferred sequence of operations for every machine in which GA is an indirect method. After that various efforts have been made to adapt genetic algorithms to solve different JSPs and have been improving the performance of genetic search by integrating other heuristic methods. Lien and Huang [6] reviewed the studies on solving JSP problems by GAs. Rudolph [7] studied the GA and proposed a method to generate a good initial population by evolving priority dispatching rules to the arrival of optimal final solutions. Masato et al. [8] put forward a genetic algorithm with modified crossover operator and search area adaptation for the job-shop scheduling problem and compared GA and SA showing that the former is more efficient than the latter. Zhang and Chong [9] proposed a multi-objective genetic algorithm incorporated with two problem-specific local improvement strategies to solve a bi-objective optimization problem. Xing et al. [10] took into account the shortest processing time and the balanced use of machines and put forward the multi-population genetic algorithm based on the multi-objective scheduling of flexible job-shop.

Particle swarm optimization is a novel evolutionary algorithm that was inspired by the motion of a flock of birds searching for foods and was proposed by Kennedy and Eberhart [11]. As a stochastic optimization method, PSO has shown good performance for solving combinatorial optimization problems, and therefore it was used to optimize JSPs [12]. But PSO's shortages are obvious, including the infirmness local convergence and the slow convergence speed at the last period. In order to get a better solution, some improved PSO algorithms were proposed. Xia et al. [13] provided a hybrid particle swarm optimization combined with simulated annealing to solve the problem of finding the minimum makespan in the job-shop scheduling environment. The simulated annealing algorithm was used to avoid becoming trapped in a local optimum. By reasonably combining these two different search algorithms, an effective hybrid optimization algorithm HPSO was developed and applied to the job-shop scheduling problem. Liang et al. [14] invented a novel PSO-based algorithm for JSPs. That algorithm can effectively exploit the capability of distributed and parallel computing systems, with simulation results showing the possibility of high-quality solutions for typical benchmark problems. Sha and Hsu [15] modified the particle position based on preference list-based representation, particle movement based on swap operator, and particle velocity based on the tabu list concept. A heuristic algorithm was used to decode a particle position into a schedule. Furthermore, they applied tabu search to improve the solution quality. Fan et al. [16] developed an improved binary particle swarm optimization (BPSO) algorithm for solving the problem of arranging m workers to process n structures, to optimize the minimum completion time of the jobs. In this improved BPSO, a new method of making initial particles was presented for searching the optimum particle in the feasible dimensional problem space. Zhang et al. [17] proposed a hybrid particle swarm optimization algorithm which was combined with a PSO algorithm and a tabu search (TS) algorithm to solve the multi-objective flexible job-shop scheduling problem (FJSP) with several conflicting and incommensurable objectives. Lin et al. [18] proposed a new hybrid swarm optimization algorithm which consisted of particle swarm optimization, simulated annealing technique and multi-type individual enhancement scheme to solve the job-shop scheduling problem. He and Jin [19] proposed a novel particle swarm optimization, which maintained the particle diversification by using the chaos mechanism, to improve the job scheduling efficiency. But these algorithms above still suffered from the problem of premature convergence, and easily trapped into local optimum. Therefore, studying new methods is still an important task to some extent.

Hormone delivery pattern to target organs, which has been demonstrated by many studies, is crucial to the effectiveness of their action. Hormone release could be altered by pathophysiology and differences in endocrine output mediate important intraspecies distinctions. Accordingly, the mechanisms controlling the dynamics of various hormones had become lately the object of extensive biomedical research [20]. In the medical field, Liu, Ren, and Ding [21] presented some simulation models of the hormone release. But there are few reports on developing the optimization algorithm by using the hormone modulation mechanism. In order to get a better self-adaptive and stable optimization algorithm, some neuroendocrine inspired optimization algorithms, which are inspired from hormone modulation mechanism, are proposed and applied to the production scheduling problem.

2.2 The Job-Shop Scheduling Problem Model

In general, the job-shop problems can be described as follows. There are a set of jobs $J = \{1, 2, ..., n\}$ and a set of machines $M = \{1, 2, ..., m\}$ to be scheduled. Each job consists of a predetermined sequence of task operations, each of which needs to be processed without pre-emption for a given period of time on a given machine. The required machine and the fixed processing time characterize each operation. There are several constraints on jobs and machines.

- Each job must visit each machine exactly once.
- There are no precedence constraints among the operations of different jobs.
- Each operation cannot be commenced until the processing is completed, if the precedent operation is still being processed.
- Tasks of the same job cannot be processed concurrently and each job must visit each machine exactly once.
- Neither release times nor due dates are specified.

A schedule is an assignment of operations to time slots on a machine. The makespan is the maximum completion time of the jobs and the objective of the JSP is to find a schedule that minimizes the makespan. A good schedule is one that minimizes the idle time the machines take. So, the problem is to determine the operation sequences on the machines in order to minimize the makespan—that is, the time from the beginning of the first task to the end of the last task, from start to finish.

The objective scheduling problem based on minimizing the maximal makespan can be formulated as follows.

$$g(x) = \min \left\{ \max_{1 \le k \le m} \left\{ \max_{1 \le i \le n} \{c_{ik}\} \right\} \right\}, \ c_{ik} \ge 0. \tag{2.1}$$

s.t:

$$c_{ik} - t_{ik} + M(1 - a_{ihk}) \ge c_{ih}, \quad a_{ihk} = 0, 1 \tag{2.2}$$

$$c_{jk} - c_{jk} + M(1 - x_{ihk}) \ge t_{jk}, \quad x_{ihk} = 0, 1 \tag{2.3}$$

where M is a very big positive number; c_{ik} denotes the finishing time of job $i(i = 1, 2, \ldots, n)$ on machine $k(k = 1, 2, \ldots, m)$; t_{ik} denotes the processing time of job i on machine k; $a_{ihk} = 1$, if job i is processed on machine h ($h = 1, 2, \ldots, m$) before on machine k, or else $a_{ihk} = 0$; $x_{ijk} = 1$, if job i is processed on machine k before job $j(j = 1, 2, \ldots, n)$, or else $x_{ijk} = 0$.

Equation (2.1) is an object function for scheduling; Eq. (2.2) presents job sequence restriction; and Eq. (2.3) presents machine restriction, namely, each machine can process only one job at a time.

2.3 Hormone Modulation Mechanism

The nervous system is an information signal system much like the nervous system. However, the nervous system uses nerves to conduct information, whereas the endocrine system mainly uses blood vessels as information channels. The nervous system is mainly responsible for the adaptive control, and the endocrine system is to regulate hormone secretion for balancing the biochemical environment of the body. In summary, the working principles of the neuroendocrine system can be concluded as neuro-control and hormone regulation.

To describe the general hormone excreting rule $F(G)$, Farhy [20] summarized the rule formulaically. The hormone excreting rule has characteristics with monotone and nonnegative, and the hormone modulation function $F_{up}(G)/F_{down}(G)$ to comply with Hill function, as shown in Eqs. (2.4) and (2.5).

$$F_{up}(G) = \frac{(G/T)^n}{1 + (G/T)^n} \tag{2.4}$$

$$F_{\text{down}}(G) = \frac{1}{1 + (G/T)^n} \tag{2.5}$$

where $T > 0$ is called a threshold; $n \geq 1$ is called a Hill coefficient; G is an independent variable.

If the excreting speed S_x of hormone x is controlled by the concentration C_y of hormone y, using Eqs. (2.4) and (2.5), then the term controlling the secretion of hormone x in the form is shown in Eq. (2.6).

$$S_x = aF(C_y) + S_{x0} \tag{2.6}$$

where a is a constant, S_{x0} is independent of hormone x and controls the initial excreting speed of hormone x.

An inspiration is obtained from the hormone modulation mechanism, and then some neuroendocrine-inspired optimization algorithms are designed.

2.4 An IAPSO for Job-Shop Scheduling Problem

Generally, the traditional PSO comprises the basic optimization concept that individuals are evolved by cooperation and competition among the individuals to accomplish a common goal. In the search process, each particle of the swarm shares the mutual information globally and benefits from the discoveries and previous experiences of all other colleagues. The hormone modulation mechanism works in the biological body in a similar way. The control performance of the hormone modulation mechanism contacts each organ in the biological body and coordinates them, and the quick self-adaptation and self-organization process on the changes of inside and outside environment is realized by moderation and concurrent control among them. An inspiration is obtained from the hormone modulation mechanism, and then an adaptive hormonal factor (HF) is designed, and therefore an improved adaptive particle swarm optimization algorithm (IAPSO) is proposed to handle JSP for better searching efficiency and quality [21].

2.4.1 Traditional PSO

Inspired by the social behavior of animals such as bird flocking, Kennedy and Eberhart proposed a novel evolutionary computation technique called Particle swarm optimization (PSO). PSO is initialized with a population (named swarm in PSO) of random solutions. Each individual or potential solution, named particle, flies in the D-dimensional problem space with a velocity that is dynamically adjusted according to the flying experiences of its own and its colleagues. It possesses the properties of easy implementation and fast convergence. Suppose that there are n particles in the L-dimensional searching space, the standard algorithm is given in some form resembling the following [22].

$$V_i(k+1) = \omega \times V_i(k) + C1 \times rand1() \times (X_{pbesti}(k) - X_i(k))$$
$$+ C2 \times rand2() \times (X_{gbesti}(k) - X_i(k)) \tag{2.7}$$

$$X_i(k+1) = V_i(k+1) + X_i(k) \tag{2.8}$$

where $V_i(k)$, called the velocity for particle i, represents the distance to be traveled by this particle from its current position, $X_i(k)$ represents the particle position. P_{best}, which denotes the personal best position (local best solution), represents ith particle's best previous position, and G_{best}, which denotes the global best position, represents the best position among all particles in the swarm. The inertia weight w is used to coordinate the flying velocity of the particle between global exploration and local exploitation. The acceleration constants $C1$ and $C2$ are two learning factors, which are used to control the flying direction of the particle. The $rand1()$ and $rand2()$ are the random variables following the uniform distribution between [0, 1].

For Eq. (2.7), the first part represents the inertia of the previous velocity. The second part is the "cognition" part, which represents the private thinking by itself. The third part is the "social" part, which represents the cooperation among the particles. The process for implementing the PSO algorithm is as follows:

(1) Initialize a swarm of particles with random positions and velocities in the D-dimensional problem space.
(2) For each particle, evaluate the desired optimization fitness function.
(3) Compare particle's fitness value with particle's pbest. If the current value is better than pbest, then set pbest value equal to the current value, and the pbest position equal to the current position in D-dimensional space.
(4) Compare fitness evaluation value with the swarm's overall previous best. If the current value is better than gbest, then reset gbest to the current particle's value.
(5) Change the velocity and position of the particle according to Eqs. (2.7) and (2.8), respectively.
(6) Loop to step 2 until the termination criterion is met, usually a sufficiently good fitness value or a specified number of generations.

In PSO, each particle of the swarm shares mutual information globally and benefits from the discoveries and previous experiences of all other colleagues during the search process. So the PSO should be effective in solving practical optimization problems. The traditional PSO design is suited to a continuous solution space, and it is difficult to address combinatorial optimization problems without modification. For better solving combinatorial optimization problems, IAPSO is proposed, which is inspired by hormone modulation mechanism, to handle the job-shop scheduling problem.

2.4.2 IAPSO Based on the Hormone Regulation Mechanism

2.4.2.1 Encoding and Decoding

One of the key issues in applying PSO successfully to JSP is how to encode a schedule to a search solution, i.e., finding a suitable mapping between problem solution and PSO particle. The encoding of PSO for JSP in the manufacturing system can be classified as direct encoding and indirect encodings. The direct encoding includes job-based encoding, operation-based encoding, finishing time-based encoding, and random key-based encoding, and so on. The indirect encoding includes priority rule-based encoding, machine-based encoding, and so on.

In this article, the operation-based representation method, which uses the integer encoding, is adopted to encode a schedule as a sequence of jobs and operations. Taking a three-job and three-machine, for example, to explain the operation-based encoding and decoding method, its technological restriction is presented in Table 2.1.

An L-dimensional particle's position vector is constructed, where L is the total number of the operations. Each dimension in a single particle denotes each operation of a job, and a particle can be decoded into a schedule indirectly. For a scheduling problem containing three jobs and each job consisting of three operations, a nine-dimensional particle vector $X(L)$ is defined in Fig. 2.1.

As shown in Fig. 2.1, A nine-dimensional chromosome vector OP as a schedule is constructed: $OP = [1, 1, 1, 2, 2, 2, 3, 3, 3] = [op_{211}, op_{111}, op_{122}, op_{312}, op_{223}, op_{321}, op_{133}, op_{333}, op_{232}]$, where op_{ijk} denotes that the jth operation of job i on machine k. The first dimension "2" of the position vector $X(L)$ is the first operation of job2,

Table 2.1 A 3 × 3 JSP problem

Job	Operations		
(a) *Operation index*			
Job1	o_{11}	o_{12}	o_{13}
Job2	o_{21}	o_{22}	o_{23}
Job3	o_{31}	o_{32}	o_{33}
(b) *Machine and time*			
Operation	Machine	Time	
o_{11}	1	2	
o_{12}	3	3	
o_{13}	2	3	
o_{21}	2	2	
o_{22}	3	5	
o_{23}	1	4	
o_{31}	2	3	
o_{32}	1	2	
o_{33}	3	3	

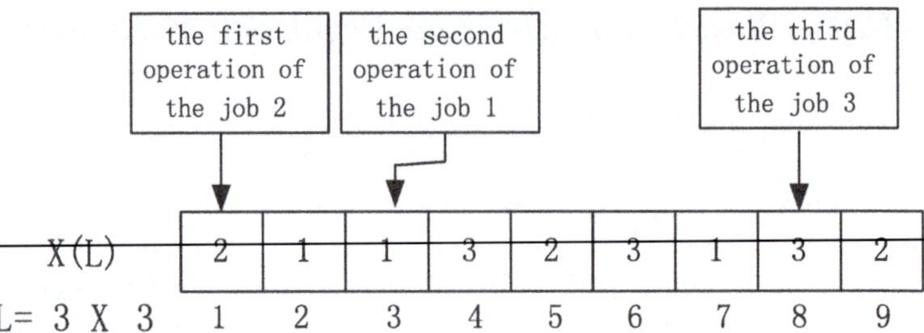

Fig. 2.1 Particle encoding

which is performed on machine 1 and the processing time 1. The second dimension "1" of $X(L)$ is the first operation of job1, which is performed on machine 1 and the processing time 3. The third dimension "1" of $X(L)$ is the second operation of job1, which is performed on machine 2 and the processing time 3. The fourth dimension "3" of $X(L)$ is the first operation of job3. The fifth dimension "2" of $X(L)$ is the second operation of job2, and so on. Therefore, the scheduling result is obtained (Fig. 2.2).

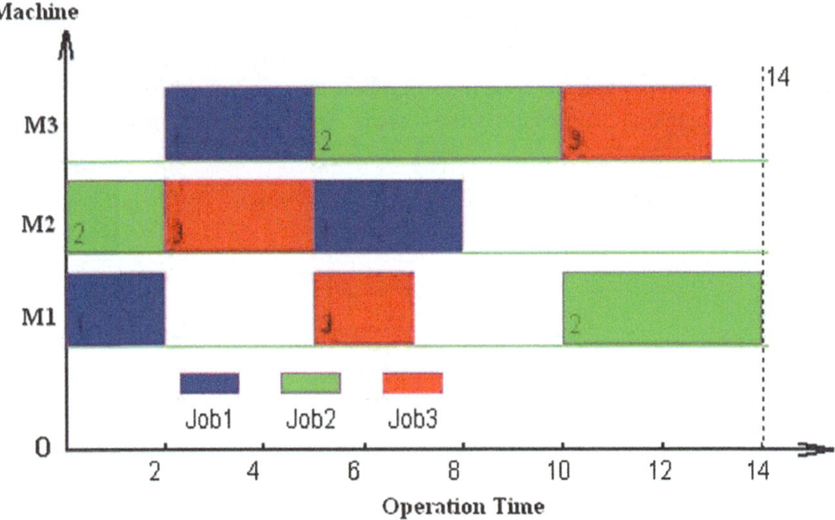

Fig. 2.2 Gantt diagram obtained by operation-based encoding and decoding

2.4.2.2   Initial Population

A PSO must be initialized with a starting population. Initial swarm and initial particle velocities are generated randomly. The encoding scheme mentioned above is a convenient tool to reproduce the initial population by using the swap-change method. The swap-change method randomly changes the positions of jobs, and does not produce the conflict problem of precedence. The initialization procedure produces N feasible particles, where N denotes the particle swarm size, by the following algorithm:

Step 1: Create an initial particle vector OP.

Step 2: Generate two integer values *rand*1 and *rand*2, by random, where both *rand*1 and *rand*2 belong to $[1, m \times n]$, where m is the number of machines and n is the number of jobs. Execute the swap-change in two genes of a particle vector OP at the positions *rand*1 and *rand*2. For example, if $OP = (3\ \underline{1}\ 3\ 1\ \underline{2}\ 2)$, *rand*1 = 2, *rand*2 = 5, then $OP' = (3\ \underline{2}\ 3\ 1\ \underline{1}\ 2)$.

Step 3: Repeat $m \times n$ times in Step 2 and then produce a new particle OP.

Step 4: Repeat steps 1–3 above N times and then produce N initial feasible solutions.

2.4.2.3   Fitness Function

Fitness is used as performance evaluation of particles in the swarm. Fitness is usually represented with a function $g: S \times R^*$, where S is the set of candidate schedules, and R^* is the set of positive real value. Mapping an original objective function value to a fitness value that represents the relative superiority of particles is a feature of the evaluation function. Generally speaking, the objective function is selected as the fitness function. In JSP, the objective function is to minimize the maximum of complete-time on all machines or other cost functions. Therefore, the reciprocal of the maximal makespan (mentioned in Eq. (2.1)) is selected as the fitness function, and the fitness of each particle is calculated according to Eq. (2.9).

$$f(x) = \frac{1}{g(x)} \tag{2.9}$$

The higher the fitness value is, the better the particle position in the search space is. In each generation, the particle with the highest fitness in the swarm is the *gbest*. The result we desired is the schedule decoded from the *gbest* of the final generation.

2.4.2.4   Adaptive Modulation Factor *HF*

As we all know, PSO has a good performance for solving combinatorial optimization problems, but the PSO algorithm also has the limitations of poor convergence, and is easy to fall in local optima. According to the Eqs. (2.7) and (2.8), the new position $X(k + 1)$ of the particle i is decided by tracing the present position $X(k)$ and two "extreme point" which are the personal best solution *pbesti* and the global best solution *gbesti*. In the traditional PSO, each particle is individual, and it becomes the traditional PSO's shortcoming that the single particle has no relations with each other. In order to get a better solution, an adaptive hormonal factor (HF) is designed by referring to Eq. (2.6) as follows.

$$HF_i = H_{\text{local}} * H_{\text{global}} \tag{2.10}$$

where H_{local} is the adaptive local hormonal factor decided by the Eq. (2.11), and H_{global} is the adaptive global hormonal factor decided by the Eq. (2.12).

$$H_{\text{local}} = a \tan\left(f_i - \frac{f_{i-1} + f_{i+1}}{2} \right) + \frac{\pi}{2} \tag{2.11}$$

$$H_{\text{global}} = a \tan\left(\frac{f_{\max} - f_{\text{avg}}}{f_{\max} - f_i} \right) \tag{2.12}$$

In Eqs. (2.11) and (2.12), $f_{i-1}, f_i,$ and f_{i+1} represent the fitness value of the particle $i-1, i,$ and $i + 1$; f_{avg} and $f_{\max}$ represent the average fitness and maximal fitness of the individual of each generation, respectively. If the particle's present fitness $f_i > (f_{i-1} + f_{i+1})/2$, which represents a high local performance of the particle i, the value of H_{local} should be updated within little fluctuation; in contrast, if $f_i < (f_{i-1} + f_{i+1})/2$, the value of H_{local} should be increased. At the same time, if the particle's present fitness $f_i > f_{\text{avg}}$ which represents a high performance of the particle i, the hormonal secretion (H_{global}) should be decreased. But if $f_i < f_{\text{avg}}$, the hormonal secretion (H_{global}) should be increased. Therefore, according to Eqs. (2.7), (2.8) and (2.10), the position of particles can be updated by the following equation:

$$X_i(k + 1) = \omega \times V_i(k) + C1 \times rand1() \times (X_{pbesti}(k) - X_i(k))$$
$$+ C2 \times rand2() \times (X_{gbesti}(k) - X_i(k)) + X_k(k) + HR \tag{2.13}$$

2.4.2.5 Algorithm Procedure

The whole IAPSO algorithm procedure is presented below.

Step 1: Initialize parameters, including swarm size S, maximum iteration K; initialize a swarm of particles (the size of each particle is $m \times n$) with positions X_i ($1 \leq i \leq s$) by initial particles generation mechanism; initialize velocities v_i; and $X_{pbesti} = X_i$.

Step 2: Evaluate each particle's fitness by calculating the value of objective function Eq. (2.1) for each particle. Initialize X_{gbest} position with the particle with the lowest fitness in the swarm.

Step 3: Set iteration number k to 1.

Step 4: If k is less than or equals K, generate next swarm using Eqs. (2.7) and (2.13) according to the method introduced above. To update the velocities of particles, we set $w = w_{max} - k(w_{max} - w_{min})/K$ and $C1 = C2 = 2.0$.

Step 5: Compute each particle's fitness and update the personal best solution *pbest*.

Step 6: Update the global best solution *gbest*.

Step 7: Set $k = k + 1$, and loop to step 4.

Step 8: Output the optimization results, and the algorithm is complete.

2.5 An IAGA for Job-Shop Scheduling Problem

2.5.1 Traditional GA

GA is stochastic search techniques based on analogy to Darwinian natural selection. Individuals who fit the environment best should have a better chance to propagate their offspring. For the same reason, solutions that have the best "fitness" should receive a higher probability to search their "neighbors". The main advantage of GA lies in its powerful implicit parallelism. In Holland's theory, a GA implicitly evaluates a number of patterns larger than population size without additional computational time and memory. The overall structure of GA can be described as follows:

(1) Encoding: For any GA, a chromosome encoding is needed to describe each chromosome in the population. The encoding method determines how the problem is structured in the algorithm and the genetic operators that are used. Each chromosome is made up of a sequence of genes from a certain alphabet which can consist of binary digits (0 and 1), floating-point numbers, integers, symbols (i.e., A, B, C, D), etc. Each chromosome represents a solution to the problem.

(2) Initial population: An initial population of size P can be randomly generated. The length of each chromosome in a population should be the same.

(3) Fitness evaluation: The fitness is computed for each chromosome in the current generation.

(4) Selection: At each iteration, the best chromosomes are chosen for reproduction by one among three different methods, i.e., binary tournament, n-size tournament, and linear ranking.

(5) Offspring generation: The offspring generation is obtained by the processes of crossover and mutation. New individuals are generated until a fixed maximum number of individuals is reached.

(6) Termination criterion: A fixed number of generations is reached. If the stop criterion is satisfied, the algorithm ends and the best chromosome, together with the corresponding result, is given as output. Otherwise, the algorithm iterates again steps 3–5.

The basic flowchart of traditional GA is presented in Fig. 2.3. However, the traditional GA is not easy to regulate GA's convergence so that GA often suffers from

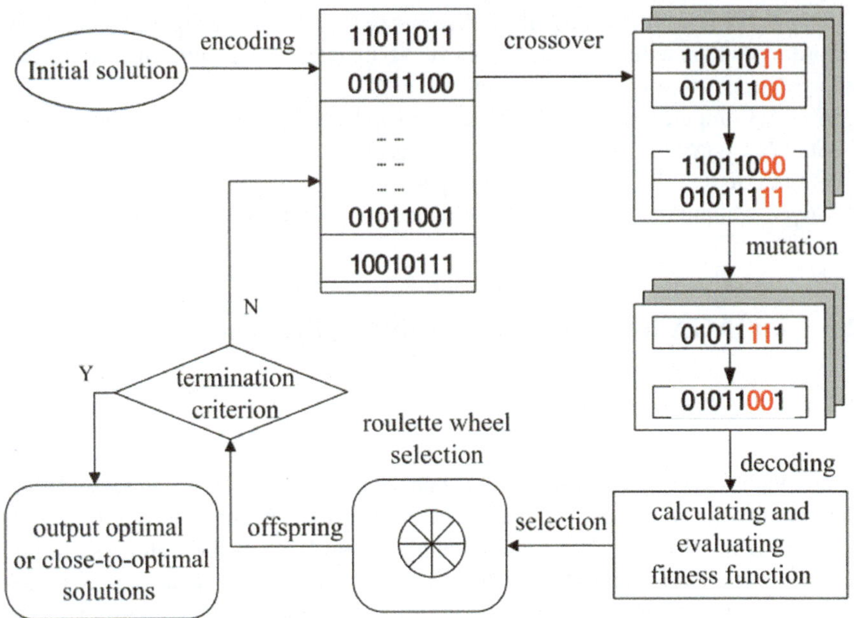

Fig. 2.3 Basic flowchart of GA

premature convergence. What's more, its crossover probability and mutation probability are fixed, the self-adaptation is not enough good. Therefore, in order to improve the disadvantages of the traditional GAs, an IAGA based on hormone modulation mechanism is proposed to apply to JSP [23].

2.5.2 An IAGA for Job-Shop Scheduling Problem

2.5.2.1 Encoding and Decoding

The encoding of IAGA for JSP in the manufacturing system can be classified as direct encoding and indirect encoding. The direct encoding includes job-based encoding, operation-based encoding, finishing time-based encoding, and random key-based encoding, and so on. Indirect encoding includes priority rule-based encoding, machine-based encoding, and so on. For example, if an operation-based encoding is adopted, then the arrangement of job represents a chromosome and represents a feasible solution.

2.5.2.2   Initial Population

An IAGA must be initialized with a starting population. The encoding scheme mentioned above is a convenient tool to reproduce the initial population by using the swap-change method.

The swap-change method randomly changes the positions of jobs, and does not produce the conflict problem of precedence. The initialization procedure produces N feasible chromosome, where N denotes the chromosome population size, by the following algorithm:

Step 1: Create an initial chromosome vector OP.

Step 2: Generate two integer values r_1 and r_2, by random, where both r_1 and r_2 belong to $[1, m \times n]$, where m is the number of machines and n is the number of jobs. Execute the swap-change in two genes of a chromosome vector OP at the positions r_1 and r_2.

Step 3: Repeat $m \times n$ times in Step 2 and then produce a new chromosome OP.

Step 4: Repeat Steps 1–3 above N times and then produce N initial feasible solutions.

2.5.2.3   Fitness Function Decision

Generally speaking, the objective function is selected as the fitness function. The reciprocal of the maximal makespan is selected as the fitness function, and then the fitness of each chromosome can be calculated according to Eq. (2.14).

$$g(x) = 1/f(x) \tag{2.14}$$

2.5.2.4   Selection

The best solution should be considered in subsequent generations. At a minimum, the single best solution from the parent generation needs to be copied to the next generation thus ensuring that the best score of the next generation is at least as good as the prior generation. Here the selection probability R is adopted according to Eq. (2.15). For example, probability 1% means that we clone the top 1% of the population solutions for the next generation.

$$R = \frac{g_i}{\sum g_i} \tag{2.15}$$

where g_i is individual fitness, $\sum g_i$ is the sum of the individual fitness.

If the number of population is N, then the number of each chromosome for the next generation is given as follows.

$$N_i = R \cdot N = \frac{g_i}{\sum g_i} \cdot N \tag{2.16}$$

In this proposed algorithm, the individual which has the highest fitness is selected directly as the next individual; therefore, the best individual can be well propagated.

2.5.2.5 Crossover

Generally speaking, crossover is the breeding of two parents to produce a single child. That child has features from both parents and thus may be better or worse than either parent according to the objective function. The traditional crossover operator (such as partly matched crossover, order crossover, and cycle crossover) would make some unfeasible solution to be created; therefore, in order to ensure to create a feasible solution, a new method for crossover operation is adopted, named as, parthenogenetic operation (PGO). The crossover operation procedure is as follows:

Step 1: A positive integer r is created randomly, and r is the position of the rth gene.

Step 2: The gene positions behind the rth gene position are moved to the front.

Step 3: The remainder gene positions are moved to the back.

For example, if a chromosome $C_1 = (\underline{3}\ \underline{1}\ \underline{5}\ 4\ 2\ 6)$ and single-point crossover $r = 3$, then $C_1' = (4\ 2\ 6\ \underline{3}\ \underline{1}\ \underline{5})$. Therefore, the unfeasible solution is avoided.

The fitness of the new individual created is examined all the time in the crossover process. If the fitness of the new individual is lower than its parent generation, then it is deleted and replaced by the individual of the parent generation. It is not finished until all the individuals of the parent generation are selected as father generation a time.

In order to accelerate the evolutional speed and enlarge the searching scope, an adaptive crossover probability p_c is designed by referring to Eq. (2.6) as follows.

$$p_c = 1 - p_c^0 \left(1 + \alpha \frac{(g_{av})^{n_c}}{(g_{max} - g_{min})^{n_c} + (g_{av})^{n_c}} \right) \tag{2.17}$$

In Eq. (2.17), p_c^0 and p_c represent initial crossover probability and adaptive crossover probability, respectively; g_{av}, g_{max}, and g_{min} represent the average fitness, maximal fitness and minimal fitness of the individual of each generation, respectively; a and n_c are coefficient factors. If the average fitness becomes big, then the crossover probability also becomes big, and vice versa.

2.5.2.6 Mutation

The primary purpose of mutation is to introduce variation and help bring back some essential genetic traits, and also to avoid the premature convergence of the entire feasible space caused by some super chromosomes. The traditional mutation procedure of the GA is to exchange the positions of a number of randomly selected genes or change the randomly selected genes to different ones. However, for the distributed manufacturing system, these mutating methods may create some unfeasible solutions. To resolve this problem, we exchange the positions of two genes for mutation operation in the proposed IAGA. The procedures are depicted as follows.

Step 1. Two positive integers r_1 and r_2 are selected randomly, and r_1 is not equal to r_2.

Step 2. Exchange the value of the position of the two genes.

For example, if $C_1 = (3\,\underline{1}\,5\,4\,\underline{2}\,6)$, $r_1 = 2$, $r_2 = 5$, then $C_1' = (3\,\underline{2}\,5\,4\,\underline{1}\,6)$.

The fitness of the new individual created is examined in the mutating process as well. If the fitness of the new individual is lower than its parent generation', then it is deleted and replaced by the individual of the parent generation. In order to accelerate the convergent speed and increase individual diversities, an adaptive mutation probability pm is also designed by referring to Eq. (2.6) as follows:

$$p_m = p_m^0 \left(1 + \beta \frac{(g_{\mathrm{av}})^{n_m}}{(g_{\max} - g_{\min})^{n_m} + (g_{\mathrm{av}})^{n_m}} \right) \tag{2.18}$$

In Eq. (2.18), p_m^0 and p_m represent initial mutation probability and adaptive mutation probability, respectively, β and n_m are coefficient factors. The population of the next generation is made up of the new individuals created by the process of crossover and mutation in this way.

2.5.2.7 Offspring Generation

Once the chromosomes for reproduction have been selected, the crossover and mutation genetic operators are applied to produce the offspring. The offspring generation phase terminates when the maximum number of individuals in the mating pool is reached.

Steps mentioned above are repeated until the convergent condition is met or a maximal number of generations is reached, then the searching work is finished. And the best individual, together with the corresponding schedule, is given as output. The flowchart of IAGA is presented in Fig. 2.4.

2.6 Application of Neuroendocrine-Inspired Optimization Algorithms for Production Scheduling

To illustrate the effectiveness of neuroendocrine-inspired Optimization Algorithm for Production Scheduling, we provide a comprehensive experimental evaluation and comparison of the proposed neuroendocrine-inspired Optimization algorithm with other powerful methods. The neuroendocrine-inspired Optimization Algorithm has

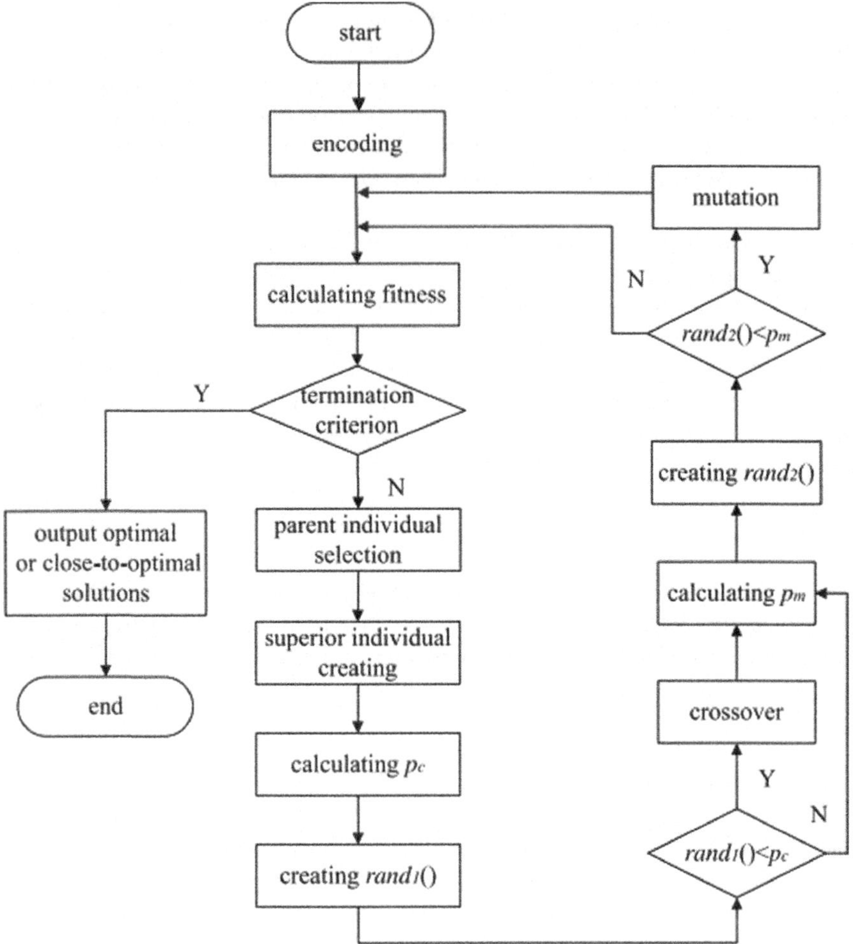

Fig. 2.4 Flowchart of IAGA

been implemented by using ANSI C programming language with the environment of Microsoft Visual C++ 6.0, and simulated it with a 1.73 GHz Intel Pentium M PC. Our experiments are done on a 1.73 GHz Intel Pentium M personal computer (PC) with 2 GB memory under Windows 10.

2.6.1 Application of the IAPSO for the JSP

To verify the performance of IAPSO for JSP, we provide a comprehensive exper- imental evaluation and comparison of the proposed IAPSO algorithm with other powerful methods. The well-known standard benchmark set of Lawerence [24] is used. To compare the performance of the IAPSO algorithm with known techniques from the literature, we compared it with the traditional PSO, TSSB [25], and HGA [26].

The following example taken from a real factory environment illustrates the application of IAPSO in the manufacturing system. There are 10 jobs that need processing on six machines, it is a technological restriction and is presented in Table 2.2.

Each row of the table represents a processing sequence of a job; for example, the numbers "$j1$ 1 (1, 6) 2 (2, 3) 3 (5, 2) 4 (3, 8) 5 (7, 3) 6 (6, 3) 7 (4, 4) 8 (9, 8) 9 (8, 2) 10 (10, 6)" in the second row of the table indicate that the working procedures of $j1$ are processed, respectively, on machines 1, 2, 5, 3, 7, 6, 4, 9, 8, and 10, and the corresponding processing time is 6, 3, 2, 8, 3, 3, 4, 8, 2, 6 in the second row, namely the processing time of the first sequence of $j1$ on machine 1 is 6 time units, the processing time of the second sequence of $j1$ on machine 2 is 3 time units, and so on.

The parameters used during the experimental process in Eqs. (2.7) and (2.8Õ) are defined in the following. Set swarm size S is 60, maximum evolutional generation iteration K is 300; other parameters are $C_1 = C_2 = 2.0$, $\omega_{max} = 1.2$, $\omega_{min} = 0.2$, respectively, what's more, these parameters (C_1, C_1, ω_{max} and ω_{min}) are well selected by more experiments. The scheduling Gantt diagram is shown in Fig. 2.5. The optimal value (92 time units) can be obtained by using PSO and IAPSO. It needs only 158 evolutional generation numbers by using IAPSO for search. What's more, this result is obtained by a random experiment. However, the best result which is selected from more experiments needs 231 evolutional generation numbers for search by using traditional PSO. The result indicates that the convergence rate of IAPSO is remarkably better than that of the traditional PSO.

To get the average performance of IAPSO and the traditional PSO, we run each instance twenty times and the solution quality was averaged. Regarding the performance measures, the average relative percentage deviation (ARD) of each instance is computed as follows:

$$\text{ARD} = \frac{\sum_{i=1}^{N}\left(\frac{f_{besti} - \text{BKS}}{\text{BKS}} \times 100\%\right)}{N} \tag{2.19}$$

where N is the run times, f_{besti} is the makespan obtained for the ith running of an algorithm and BKS is the known minimum makespan for the problem or the best-known solution for Lawerence's instances. The results of the benchmark problems for all the algorithms are provided in Table 2.3.

In Table 2.3, "Instance" means the problem name, "Size" means the problem size n jobs on m machines, and the boldface represents the better solution for one instance that at least one of the four algorithms cannot obtain the best-known solution, respectively. According to Table 2.3, IAPSO can find the best-known solution with 37 instances, and the other results of LA24, LA29 and LA40 are also much better than PSO, TSSB, and HGA. This fact shows that the IAPSO could obtain better solutions than the other algorithms.

Table 2.2 The processing information of the test example

Job	Sequence (machine number, processing time)									
j1	1 (1, 6)	2 (2, 3)	3 (5, 2)	4 (3, 8)	5 (7, 3)	6 (6, 3)	7 (4, 4)	8 (9, 8)	9 (8, 2)	10 (10, 6)
j2	1 (3, 1)	2 (6, 5)	3 (4, 3)	4 (10, 8)	5 (8, 6)	6 (2, 11)	7 (7, 2)	8 (1, 3)	9 (9, 5)	10 (5, 9)
j3	1 (2, 11)	2 (1, 3)	3 (9, 7)	4 (3, 2)	5 (10, 10)	6 (7, 2)	7 (8, 3)	8 (6, 4)	9 (5, 4)	10 (4, 7)
j4	1 (4, 3)	2 (6, 5)	3 (8, 12)	4 (7, 5)	5 (1, 3)	6 (9, 3)	7 (2, 8)	8 (3, 4)	9 (5, 12)	10 (10, 9)
j5	1 (7, 2)	2 (6, 6)	3 (10, 10)	4 (9, 6)	5 (5, 1)	6 (1, 1)	7 (2, 2)	8 (3, 8)	9 (4, 3)	10 (8, 7)
j6	1 (3, 3)	2 (8, 9)	3 (9, 4)	4 (4, 6)	5 (6, 4)	6 (7, 2)	7 (1, 2)	8 (2, 3)	9 (5, 2)	10 (10, 9)
j7	1 (3, 3)	2 (7, 1)	3 (4, 12)	4 (9, 4)	5 (10, 2)	6 (1, 8)	7 (2, 3)	8 (5, 2)	9 (6, 8)	10 (8, 7)
j8	1 (3, 3)	2 (4, 5)	3 (10, 4)	4 (8, 3)	5 (6, 4)	6 (5, 10)	7 (2, 5)	8 (7, 2)	9 (1, 3)	10 (9, 11)
j9	1 (10, 8)	2 (7, 1)	3 (4, 12)	4 (9, 4)	5 (3, 2)	6 (8, 8)	7 (2, 3)	8 (3, 8)	9 (5, 8)	10 (1, 7)
j10	1 (3, 6)	2 (10, 5)	3 (4, 4)	4 (8, 3)	5 (6, 4)	6 (5, 10)	7 (9, 5)	8 (7, 6)	9 (1, 3)	10 (2, 11)

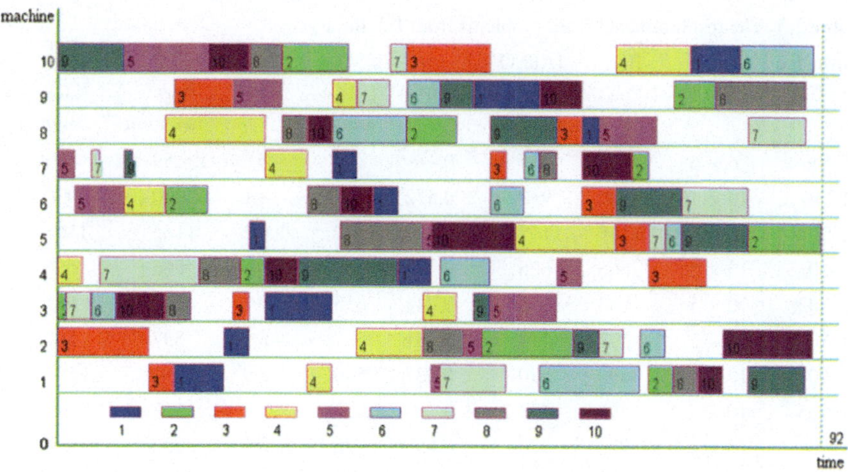

Fig. 2.5 Scheduling Gantt diagram

The comparison of the average relative percentage deviation between the IAPSO and PSO from the thirty instances is shown in Fig. 2.6. From Fig. 2.6, the first fact shows that IAPSO can provide a better searching ability than the traditional PSO. In general, the PSO is easy to be trapped in a local optimal and cannot find a better solution. From the results of Table 2.3 and Fig. 2.6, using an adaptive hormonal factor (*HF*), IAPSO effectively increases the local searching ability of the original PSO for the JSP scheduling problems. Observing the curve diagram in Fig. 2.6, the second fact is that the difference between the IAPSO's Best solution and the BKS and the IAPSO's ARD are all within 3.3%. This fact shows that the solution of IAPSO is quite stable.

The average computation time on the test problems in CPU seconds is shown in Table 2.4. The "BST" denotes the average computation time that the algorithm takes to find the best solution. Form Table 2.4, we can see that the average "Best solution time" of the proposed IAPSO is less than the computation time of the traditional PSO. It means that the IAPSO could find the optimization schedule more efficiently than the traditional PSO.

Based on the above experiments, we can obtain that, by using the adaptive hormonal factor (*HF*), each particle of the swarm shares the more mutual information globally, and the better solution can be obtained in the search space. Another advantage of the IAPSO algorithm is that it is much simpler and easy to be implemented.

Table 2.3 The result of benchmark problems from JSP library

Instance	Size ($n \times m$)	BKS	IAPSO		PSO		TSSB	HGA
			Best solution	ARD	Best solution	ARD	Best solution	Best solution
FT06	6×6	55	55	0	55	0	55	55
FT10	10×10	930	930	0.572	1007	8.1	930	930
FT20	20×5	1165	1165	0.327	1242	6.2	1165	1165
LA01	10×5	666	666	0	666	0	666	666
LA02	10×5	655	655	0.244	655	7.1	655	655
LA03	10×5	597	597	0.381	597	5.3	597	597
LA04	10×5	590	590	0.537	590	2.9	590	590
LA05	10×5	593	593	0	593	0	593	593
LA06	15×5	926	926	0.453	926	3.1	926	926
LA07	15×5	890	890	0	890	0	890	890
LA08	15×5	863	863	0	863	0	863	863
LA09	15×5	951	951	0	951	0	951	951
LA10	15×5	958	958	0	958	0	958	958
LA11	20×5	1222	1222	0	1222	0	1222	1222
LA12	20×5	1039	1039	1.299	1039	4.6	1039	1039
LA13	20×5	1150	1150	0.941	1150	3.6	1150	1150
LA14	20×5	1292	1292	0	1292	0	1292	1292
LA15	20×5	1207	1207	0	1207	0	1207	1207
LA16	10×10	945	945	1.248	945	14.8	945	945
LA17	10×10	784	784	0.127	784	3.1	784	784
LA18	10×10	848	848	1.005	848	8.8	848	848
LA19	10×10	842	842	0.772	842	9.5	842	842
LA20	10×10	902	902	1.136	907	6.3	902	907
LA21	15×10	1046	1046	3.1	1055	30.5	1046	1046
LA22	15×10	927	927	1.121	935	15.5	927	935
LA23	15×10	1032	1032	0	1032	0	1032	1032
LA24	15×10	935	**936**	3.3	938	30.8	938	953
LA25	15×10	977	977	2.6	983	22.9	979	986
LA26	20×10	1218	1218	0.640	1218	1.7	1218	1218
LA27	20×10	1235	1235	1.187	1252	17.1	1235	1256
LA28	20×10	1216	1216	1.225	1216	25.7	1216	1232
LA29	20×10	1152	**1165**	1.642	1179	36.8	1168	1196
LA30	20×10	1355	1355	0	1355	0	1355	1355
LA31	30×10	1784	1784	0.631	1784	2.3	1784	1784

(continued)

Table 2.3 (continued)

Instance	Size ($n \times m$)	BKS	IAPSO		PSO		TSSB	HGA
			Best solution	ARD	Best solution	ARD	Best solution	Best solution
LA32	30 × 10	1850	1850	2.399	1850	4.6	1850	1850
LA33	30 × 10	1719	1719	1.141	1719	3.6	1719	1719
LA34	30 × 10	1721	1721	0.513	1721	1.2	1721	1721
LA35	30 × 10	1888	1888	1.113	1888	3.1	1888	1888
LA36	15 × 15	1268	1268	3.1	1291	28.8	1268	1279
LA37	15 × 15	1397	1397	2.8	1442	33.2	1407	1408
LA38	15 × 15	1196	1196	3.2	1228	23.1	1196	1219
LA39	15 × 15	1233	1233	1.8	1233	21.6	1233	1246
LA40	15 × 15	1222	**1224**	2.3	1236	24.3	1229	1241

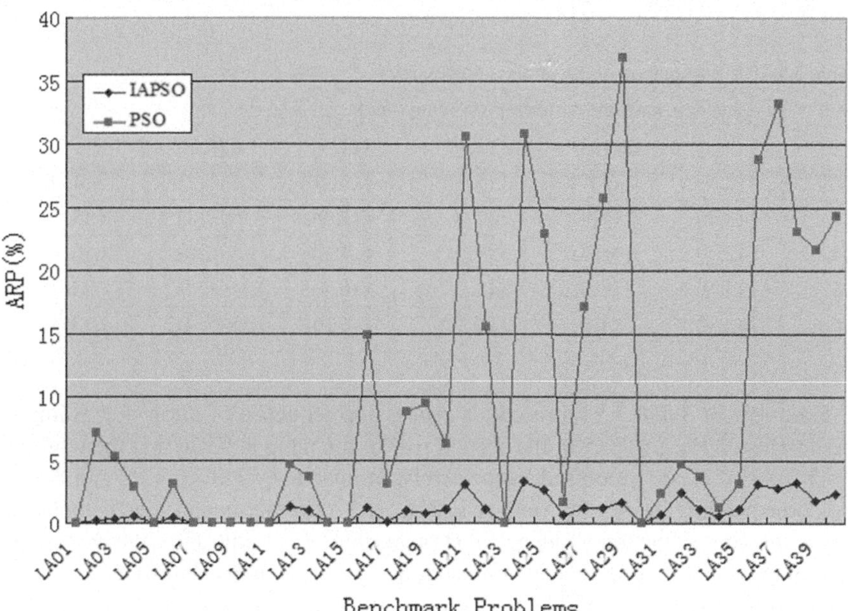

Fig. 2.6 Comparison of ARD between IAPSO and PSO

2.6.2 *The Application of the IAGA for JSSP*

The following example, taken from a real factory environment, illustrates the application of IAGA in the manufacturing system. There are six jobs that need processing on six machines, its' a technological restriction and is presented in Table 2.5.

Table 2.4 Computation time of the test problems

Problem	Size ($n \times m$)	PSO		IAPSO	
		BST	Total time	BST	Total time
FT06	6×6	0.0	32	0.0	29
FT10	10×10	21.7	91	5.3	157
FT20	20×5	19.2	138	18.8	210
LA01-05	10×5	5.3	50	0.5	37
LA06-10	15×5	0.1	92	0.3	65
LA11-15	20×5	0.5	143	0.1	98
LA16-20	10×10	15.5	90	13.1	137
LA21-25	15×10	37.2	164	28.9	170
LA26-30	20×10	103.1	259	93	383
LA31-35	30×10	31.4	520	3.0	751
LA36-40	15×15	68.4	254	65.2	463

Table 2.5 The information of the product processing

Job	Sequence (machine number, processing time)					
$j1$	1 (3, 1)	2 (1, 3)	3 (2, 6)	4 (4, 7)	5 (6, 3)	6 (5, 6)
$j2$	1 (2, 8)	2 (3, 5)	3 (5, 10)	4 (6, 10)	5 (1, 10)	6 (4, 4)
$j3$	1 (3, 5)	2 (4, 4)	3 (6, 8)	4 (1, 9)	5 (2, 1)	6 (5, 7)
$j4$	1 (2, 5)	2 (1, 5)	3 (3, 5)	4 (4, 3)	5 (5, 8)	6 (6, 6)
$j5$	1 (3, 9)	2 (2, 3)	3 (5, 5)	4 (6, 4)	5 (1, 3)	6 (4, 1)
$j6$	1 (2, 3)	2 (4, 3)	3 (6, 9)	4 (1,10)	5 (5, 4)	6 (3, 1)

Each row of Table 2.5 represents a processing sequence of a job; for example, the numbers 3, 1, 2, 4, 6, 5 in the second row of the table indicate that the working procedures of $j1$ are processed, respectively, on machines 3, 1, 2, 4, 6, and 5, and the corresponding processing time is 1, 3, 6, 7, 3, 6 in the second row, namely the processing time of the first sequence of $j1$ on machine 3 is 1 time units, the processing time of the second sequence of $j1$ on machine 1 is 3 time units, and so on.

Parameters of IAGA are selected as follows: population size N is 40, evolutional generation number G is 50, initial crossover probability p_c^0 is 0.7, and initial mutation probability p_m^0 is 0.01; other parameters are $a = 0.3$, $b = 0.2$, $n_c = n_m = 2$, respectively, what's more, these parameters (a, b, n_c and n_m) are well selected by more experiments.

Figure 2.7 shows comparisons of the convergence results using GA and IAGA under the same evolutionary environments. The scheduling Gantt diagram is shown in Fig. 2.8. The optimal value (55 time units) can be obtained by using GA and IAGA. It needs only 15 evolutional generation numbers by using IAGA for search. What's more, this result is obtained by random experiments. However, the best result

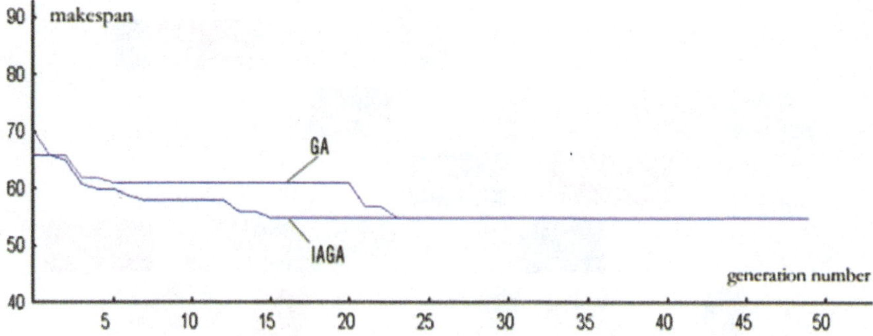

Fig. 2.7 Comparisons of some convergence results using GA and IAGA

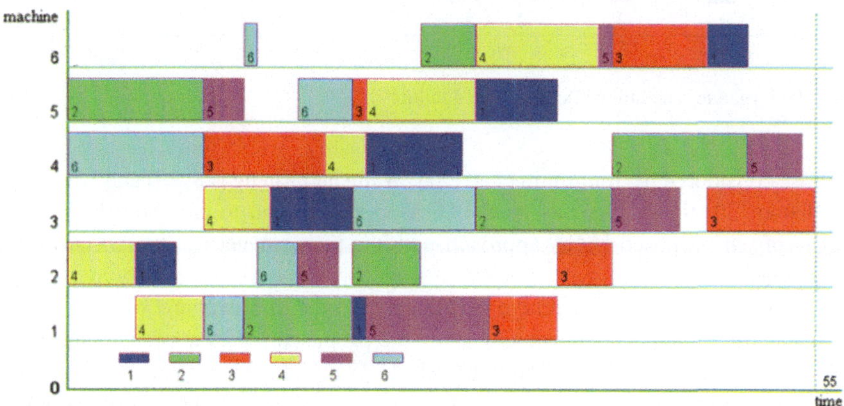

Fig. 2.8 Normal scheduling

which is selected from more experiments needs 23 evolutional generation numbers for search by using GA. The result indicates that the convergence rate of IAGA is remarkably better than that of the traditional GA.

But in a practical production process, machine failure happens randomly. Therefore, the subsequent scheduling and minimal makespan will be influenced. By using IAGA, the remainder jobs can be rescheduled dynamically and the result will be improved accordingly. Figure 2.9 shows dynamic rescheduling Gantt diagram after machine 4 is fault when the fourth operation of job1 is machined over, in order to balance utilization ratio of each machine; therefore the remainder operations (denoted by ellipse dashed) which need to be processed on machine 4 are conducted by machines 1 and 2 (supposing that each machine has an ability to process any job). An optimal scheduling result for a JSSP can be obtained quickly by our proposed IAGA as shown in Fig. 2.9. Therefore, the example reflects much better self-adaptation to JSP.

In additional, in order to compare with the other algorithms, we adopt the same example, the famous MT 10 (10 × 10) problem of Muth and Thompson benchmarks,

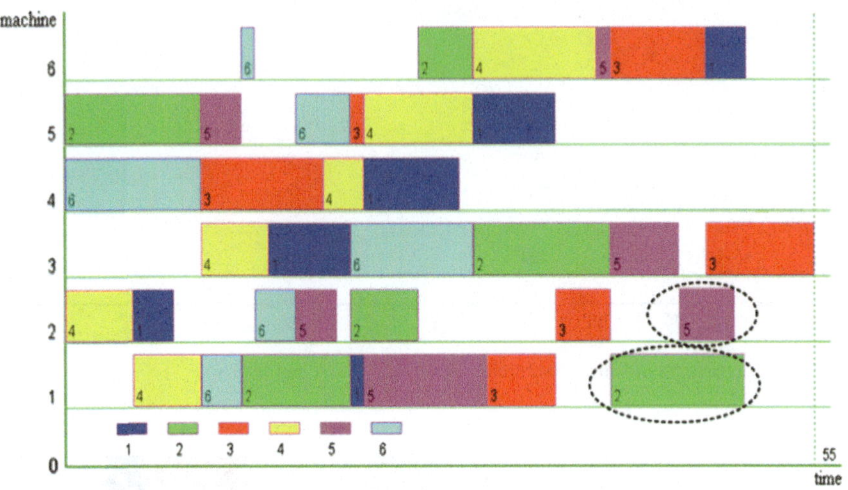

Fig. 2.9 Dynamic scheduling after machine 4 failure

as those considered by Tsujimura et al. [27], Wang and Zheng [28, 29] and Liu, Tsai, and Chou [30] to test our proposed IAGA approach, and to compare the performance of our proposed IAGA approach with the performances of those GA-based approaches.

The best makespan, the average makespan, and the improvement of the average makespan in 20 independent runs by using the proposed IAGA approach and those GA-based methods given by Tsujimura et al., Wang and Zheng and Liu et al. are shown in Table 2.6. The makespans of an optimal solution for the corresponding problem are also given in Table 2.6.

From Table 2.6, it can be seen that for MT10 (10 jobs × 10 machines) problem of Muth and Thompson benchmark, (i) the proposed IAGA approach, the HGA method given by Wang and Zheng, and the HTGA given by Liu et al. can find the optimal makespan, but the average makespan obtained by using the proposed IAGA approach is better than those obtained by using the MGA, and the HTGA methods. (ii) the proposed IAGA approach can give the better makespan and the better average makespan than those GA-based methods given by Tsujimura et al. And hence the proposed IAGA approach has a more robust and stable solution quality.

Table 2.6 Computational results using different methods for MT 10 (10 × 10) (optimal: 930)

Method	GA	SyGA1	SyGA2	GASA	HTGA	IAGA
Best	966	937	930	930	930	930
Average	993	965.85	965.5	953.7	943.48	937
Improvement (%)	5.64	2.96	2.95	1.75	0.7	0

2.7 Conclusion

The hormone modulation mechanism is one of the major physiological systems and has better self-adaptability and stability. It is a general law given by Farhy [20] and is characterized by nonnegative and monotone. In this chapter, these neuroendocrine-inspired optimization algorithms (IAPSO and IAGA), inspired by hormone modulation mechanism, have been proposed and developed for solving the production scheduling problem. In IAPSO, a new particle generation mechanism has been presented to guarantee the particles searching optimal solution in the feasible solution space efficiently, and it also shows the strong local search ability by the adaptive hormonal factor (HF). We executed the proposed IAPSO to solve a set of benchmark problems. The computational experiments show that the proposed IAPSO can find optimal or close to optimal solutions more efficiently and stably than traditional PSO.

What's more, an inspiration for GA is also obtained from hormone modulation mechanism, and then an adaptive crossover probability pc and an adaptive mutation probability pm are designed. Therefore, IAGA is proposed to solve the combinatorial optimization problems of JSP. We executed the proposed IAGA to solve a DJSP, a contrastive experiment, and a benchmark problem. The computational experiments show that the proposed IAGA can find optimal or close to optimal solutions and it is more efficient than GA [31]. What's more, the proposed algorithm is not worse than the hybrid genetic algorithm (HGA).

Therefore, the proposed neuroendocrine-inspired optimization algorithms possess the merits of global exploration, fast convergence, and robustness to solve the JSPs. And by employing these approaches, machines can be used most efficiently, which means that tasks can be allocated appropriately, production efficiency and machine usage can be improved, and the production cycle can be shortened efficiently.

In further, there are three aspects of discussion as a useful extension of this research. (1) How to enhance population diversity and improve search efficiency is the first future work. (2) The second future work is how to design a more efficient information sharing mechanism and more effective local search strategy in this proposed algorithm. (3) In order to make full use of the merits in the proposed neuroendocrine-inspired optimization algorithms and obtain the perfect results for the JSP, the hybrid search approach should be developed, which combined these proposed optimization algorithms with other search algorithms.

References

1. Brucker, P., Knust, S., Cheng, T. C., et al. (2004). Complexity results for flow-shop and open-shop scheduling problems with transportation delays. *Annals of Operations Research, 129*(1), 81–106.
2. Carey, E. L., Johnson, D. S., & Sethi, R. (1976). The complexity of flow shop and job shop scheduling. *Mathematics of Operations Research, 1*(2), 117–129.

3. Reeves, C. R. (1993). Improving the efficiency of tabu search for machine sequencing problems. *Journal of the Operational Research Society, 44*(4), 375–382.

4. Jones, D. F., & Tamiz, M. (2003). Analysis of trends in distance metric optimization modelling for operational research and soft computing. *Multi-Objective Programming and Goal Programming, 21*(3), 19–26.

5. Hsiang, P. L., & Cherng, M. W. (2008). A genetic algorithm embedded with a concise chromosome representation for job-shop scheduling problems. *Journal of Intelligent Manufacturing, 29*(1), 19–34.

6. Lien, C. C., & Huang, C. L. (1999). The model-based dynamic hand posture identification using genetic algorithm. *Machine Vision and Applications, 11*(3), 107–121.

7. Rudolph, G. (1994). Convergence properties of canonical genetic algorithms. *IEEE Transactions on Neural Networks, 5*(1), 96–101.

8. Masato, W., Kenichi, I., & Mitsuo, G. (2005). A genetic algorithm with modified crossover operator and search area adaptation for the job-shop scheduling problem. *Computers & Industrial Engineering, 48*(4), 743–752.

9. Zhang, R., & Chong, R. (2016). Solving the energy-efficient job shop scheduling problem: A multi-objective genetic algorithm with enhanced local search for minimizing the total weighted tardiness and total energy consumption. *Journal of Cleaner Production, 11*(2), 3361–3375.

10. Xing, L. N., Chen, Y. W., & Yang, K. W. (2011). Multi-population interactive coevolutionary algorithm for flexible job shop scheduling problems. *Computational Optimization and Applications, 48*(1), 139–155.

11. Kennedy, J., & Eberhart, R. C. (1995). Particle swarm optimization. In *Proceedings of the 1995 IEEE International Conference on Neural Networks* (Vol. 4, pp. 1942–1948) Piscataway, NJ: IEEE.

12. Lian, Z. G., Jiao, B., & Gu, X. S. (2006). A similar particle swarm optimization algorithm for job-shop scheduling to minimize makespan. *Applied Mathematics and Computation, 183*(2), 1008–1017.

13. Xia, W. J., Wu, Z. M., & Zhang, W. (2004). Applying particle swarm optimization to job-shop scheduling problem. *Chinese Journal of Mechanical Engineering, 17*(3), 437–441.

14. Liang, Y. C., Ge, H. W., Zhou, Y., et al. (2005). A particle swarm optimization-based algorithm for job-shop scheduling problems. *International Journal of Computational Methods, 2*(3), 419–430.

15. Sha, D. Y., & Hsu, C. Y. (2006). A hybrid particle swarm optimization for job shop scheduling problem. *Computers & Industrial Engineering, 51*(4), 791–808.

16. Fan, K., Zhang, R. Q., & Xia, G. P. (2007). Solving a class of job-shop scheduling problem based on improved BPSO algorithm. *Systems Engineering-Theory & Practice, 27*(11), 111–117.

17. Zhang, G. H., Shao, X. Y., et al. (2009). An effective hybrid particle swarm optimization algorithm for multi-objective flexible job-shop scheduling problem. *Computers & Industrial Engineering, 56*(4), 1309–1318.

18. Lin, T. L., Horng, S. J., et al. (2010). An efficient job-shop scheduling algorithm based on particle swarm optimization. *Expert Systems with Applications, 37*(3), 2629–2636.

19. He, J., & Jin, J. (2012). Research on the job shop scheduling optimization based on CPSO algorithm. *Journal of Convergence Information Technology, 7*(11), 60–66.

20. Farhy, L. S. (2004). Modelling of oscillations in endocrine networks with feedback. *Methods in Enzymology, 38*(4), 54–81.

21. Gu, W. B., Tang, D. B., & Zheng, K. (2014). Solving job-shop scheduling problem based on improved adaptive particle swarm optimization algorithm. *Transactions of Nanjing University of Aeronautics & Astronautics, 31*(2), 275–293.

22. Lei, D. M. (2008). A Pareto archive particle swarm optimization for multi-objective job shop scheduling. *Computer and Industrial Engineering, 54*(4), 960–971.

23. Wang, L., & Tang, D. B. (2011). An improved adaptive genetic algorithm based on hormone modulation mechanism for job-shop scheduling problem. *Expert Systems with Applications, 38*(6), 7243–7250.

24. Lawrence, S. (1984). An experimental investigation of heuristic scheduling techniques. In *Supplement to resource constrained project scheduling*. GSIA. Pittsburgh, PA: Carnegie Mellon University.
25. Pezzella, F., & Merelli, E. (2000). A tabu search method guided by shifting bottleneck for the job shop scheduling problem. *European Journal of Operational Research, 120*(2), 297–310.
26. Goncalves, J. F., Mendes, J. J. M., & Resende, M. G. C. (2005). A hybrid genetic algorithm for the job shop scheduling problem. *European Journal of Operational Research, 167*(1), 77–95.
27. Tsujimura, Y., Mafune, Y., & Gen, M. (2001). Effects of symbiotic evolution in genetic algorithms for job-shop scheduling. In *Proceedings of the IEEE 34th International Conference on System Sciences*, Hawaii, USA, June 2001 (Vol. 3, pp. 1–7). IEEE.
28. Wang, L., & Zheng, D. Z. (2001). An effective hybrid optimization strategy for jobshop scheduling problems. *Computers & Operations Research, 28*(6), 585–596.
29. Wang, L., & Zheng, D. Z. (2002). A modified genetic algorithm for job-shop scheduling. *Advanced Manufacturing Technology, 20*(1), 72–78.
30. Liu, T. K., Tsai, J. T., & Chou, J. H. (2006). Improved genetic algorithm for job-shop scheduling problem. *International Journal of Advanced Manufacturing Technology, 27*(9), 1021–1029.
31. Wang, W. L., Wu, Q. D., & Song, Y. (2004). Modified adaptive genetic algorithms for solving job-shop scheduling problems. *Systems Engineering-Theory and Practice, 24*(2), 58–62.

Chapter 3
Hormone Regulation Based Approach for Distributed and On-line Scheduling of Machines and AGVs

3.1 Introduction and Synopsis

With the globalization of markets, the continuous innovation of technology, and the short life of products, the traditional manufacturing industry is facing tremendous challenges. This situation has promoted the development of manufacturing systems toward the direction of discretization, intellectualization, and autonomy. A manufacturing system typically consists of several production operation units, an automated storage/retrieval system (AS/RS) for raw material and finished products, and an automated material handling system for material transfer between units. In order to assure the high agility and flexibility, manufacturing systems require an intelligent and automated material handling system [1]. An automated guided vehicle system (AGVS) is one of the most efficient material handling systems due to high flexibility, safety, and small space utilization of automated guided vehicles (AGVs). In recent years, the research on AGV technology has strengthened their flexibility, intelligence, and autonomy [2–5].

An AGVS should ensure normal and efficient material transfer, meanwhile, reducing the transportation cost, work-in-process (WIP) inventories, and total operation cost [5, 6]. To achieve these goals, many approaches and solutions are used in manufacturing systems. In these systems, scheduling problems are optimization processes of resource and task allocation. Both the scheduling of operations on machines as well as the scheduling of AGVs are essential factors contributing to the efficiency of the manufacturing systems [7]. AGV scheduling problems depend on the machine scheduling for task execution, specifically on the starting and ending time assigned by the operations' scheduling. The objective of AGV scheduling is to optimize the material handling, executed by AGVs, among the machines and the AS/RS. AGV scheduling can be achieved by both off-line and on-line approaches. Off-line scheduling is aimed to manage all activities for the entire scheduling period; while, on-line scheduling carries out the allocation and dispatching in real-time, during the execution of the plan.

© Springer Nature Singapore Pte Ltd. 2020
D. Tang et al., *Adaptive Control of Bio-Inspired Manufacturing Systems*,
Research on Intelligent Manufacturing,
https://doi.org/10.1007/978-981-15-3445-4_3

Off-line AGV scheduling is an NP-hard problem that must take into account machine scheduling and AGV scheduling simultaneously. Numerous research results show that NP-hard scheduling problems can be solved by heuristic approaches [8–10], as well as off-line AGV scheduling problem [11]. In order to optimize AGV scheduling, Confessore et al. [12] proposed a network-based heuristic approach for the vehicle dispatching. Bilge and Ulusoy [13] exploit the interactions between machine scheduling and AGV scheduling in a flexible manufacturing system (FMS) by addressing them at the same time. They define a nonlinear mixed integer-programming model to formulate the problem. They apply a heuristic-based iterative approach to generate machine schedules that are used to define time windows for AGVs; afterward a feasible solution for AGV schedules is searched. Besides, several more researches [14–17] solve scheduling problems in the literature [13] by using meta-heuristic techniques.

The goal of off-line scheduling is to generate an optimized schedule over a period of time, during which the schedule is not expected to require any change. Nevertheless, during production, disturbances are likely to cause the unavailability of the original schedule. Therefore, on-line scheduling or dispatching approaches can substantially improve the performance of AGVSs [18–20]. There are two on-line approaches for AGVSs: centralized and decentralized (or distributed). A centralized approach has one controller for all AGVs in the system [21–23]. Whereas, in a distributed approach there are several subsystems controlled by different independent controllers based on their local knowledge. Multi-Agent Systems (MASs) and Holonic Manufacturing System (HMS), with distributed characteristic, can be used to implement on-line AGV routing and dispatching. Lau and Woo [24] describe an agent-based dynamic routing strategy for generic automated material handling systems. Morten and Olivier [25] present a holonic-based approach for AGVS developed for the automated paint-shop. Babiceanu et al. [26] introduce a holonic control approach for scheduling material handling devices in the cellular manufacturing environment. There are also some researchers applying distributed approaches to deal with classic objectives existing for AGV routing and dispatching, such as deadlock, conflict, etc. Singh and Tiwari [27] propose a MAS approach for the operational control on AGV conflict-free routing. Breton et al. [28] introduce a routing approach based on MAS for deadlock avoidance and conflict free. Wallace [29] presents an agent-based AGV controller to maintain deadlock-free flow of AGVs within a free space world model.

Although there have been many researches on on-line AGV routing and dispatching by applying distributed approaches, very few carry out the study on on-line scheduling for machines and AGVs simultaneously. To the best of the authors' knowledge, only one work [30] proposes an MAS approach for the simultaneous scheduling of machines and AGVs. The approach works in an on-line environment and generates feasible schedules using negotiation and bidding mechanisms between distributed agents. MAS negotiation and biding mechanisms have the advantage of simplicity. Nevertheless, since they are a type of explicit coordination mechanism, when the complexity of the manufacturing environment increases (the number of machines and AGVs is large, together with a high number of new tasks appearing at run time), the number of messages passing is also large, and time consumption in some situations is not appropriate for given feasible solutions in due date.

Intelligent models inspired by human body information processing mechanism have become a new research focus in the Artificial Intelligence research field. For example, the endocrine system is a core part of the human body information processing mechanism. Its unique processing method, based on the principle of hormone diffusion and reaction, has given researchers a lot of inspiration [31, 32]. The hormone regulation mechanism is one kind of implicit coordination approach that leads to quick coordination between the different components of the entire system. Compared with MAS negotiation and bidding mechanisms, this mechanism implements simpler coordination and requires less message passing among the system components. In this work, an approach for distributed and on-line scheduling of AGVs and machines that uses the hormone regulation mechanism is proposed [33].

The remainder of the chapter is organized as follows. Section 3.2 presents an on-line scheduling model inspired by the principle of hormone diffusion and reaction to agilely deal with emergent tasks. Section 3.3 introduces a transportation task allocation mechanism based on the hormone regulation mechanism. Section 3.4 illustrates a hormone regulation based cooperation mechanism between machines and AGVs for on-line scheduling. Section 3.5 presents an evaluation study in which the proposed approach is compared to other state-of-the-art approaches. Section 3.6 describes the conclusions.

3.2 On-line Scheduling Model

3.2.1 On-line Scheduling Approach

An on-line problem for simultaneous scheduling of machines and AGVs can be modeled as an on-line job shop scheduling problem in which the materials are transported by AGVs. The process route plan generated from an on-line job shop scheduling determines the key time nodes of an on-line AGV scheduling. In recent years, on-line job shop scheduling problems for manufacturing systems have been studied by many researchers [34–36]. In our previous work [37], a shop floor rescheduling approach was presented. This technique can deal with on-line scheduling for machines and operations proposed in this article. Thus, the focus of this paper is the scheduling of transportation tasks between AGVs and machines, shown in Fig. 3.1.

This scheduling approach is collaborative and distributed, and can be described in an abstract way as follows. The collaborative scheduling process is divided into three stages. In stage I, the scheduling process is triggered when an operation is started on a machine; the machine must deal with the subsequent transportation throughout the shop floor before the operation is finished. The machine needs to find a suitable AGV that can execute the transportation. To this end, it sends the demand requirement to the AGVs in the system. In stage II, when a given AGV realizes the

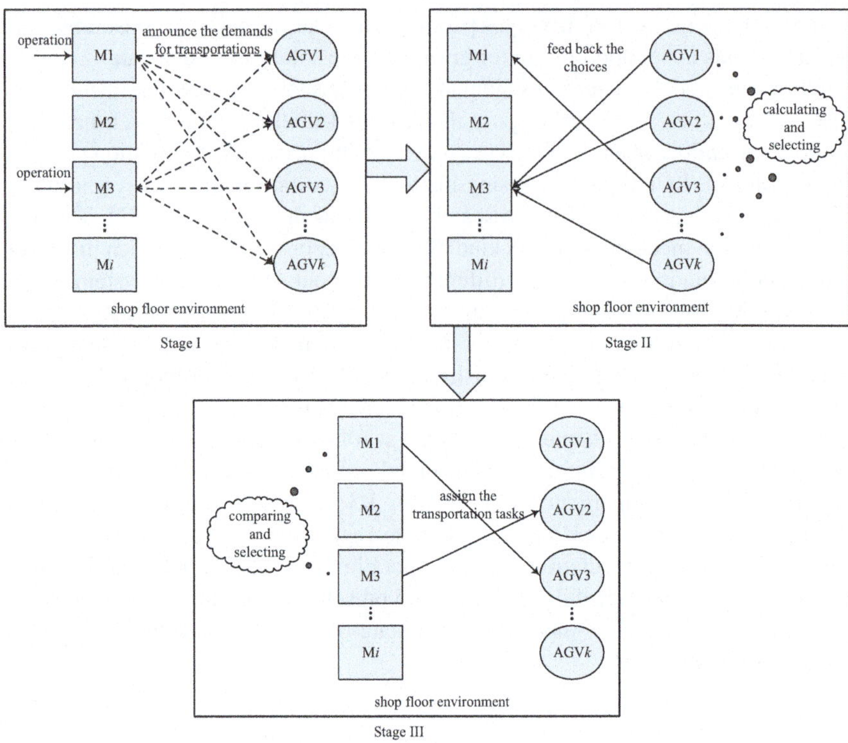

Fig. 3.1 Scheduling of transportation tasks between AGVs and machines

demands from machines (sensing the environment or receiving a message), calculates its transportation efficiency for every demand and selects the demand with the highest transportation efficiency according to its own criteria, then communicates its choice back to the relevant machine. In stage III, the machine receives/senses the feedback information, then evaluates the different alternatives and selects the most efficient one. Finally, the machine allocates the transportation task to the selected AGV. In the next subsections, an on-line scheduling model inspired by the principle of hormone diffusion and reaction is detailed.

3.2.2 *Information Processing Mechanism in Endocrine System*

The endocrine system of the human body is a complex physiology network formed by a variety of endocrine glands. These endocrine glands can influence each other by secreting, transmitting and responding to different hormones [38]. Endocrine glands with different functions have their own proprietary hormone receptors, which react

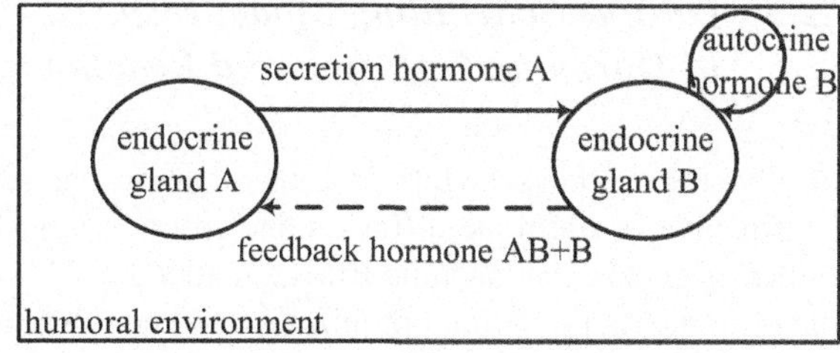

Fig. 3.2 Principle of hormone diffusion and reaction in the endocrine system

to hormones in the humoral environment. However, endocrine glands diffuse hormones into the humoral environment with no constraints; the hormone receptors only respond to the specific hormones. The principle of hormone diffusion and reaction in the endocrine system is illustrated in Fig. 3.2. The solid lines with an arrow represent the hormone secretion between endocrine glands or hormone autocrine in independent endocrine glands; whereas the dotted line with an arrow indicates the hormone feedback of endocrine glands from the participant to the initiator. For example, affected by a stimulus, endocrine gland "A" secretes hormone "A" and releases it into the humoral environment. Sensing the stimulus from hormone "A", endocrine gland "B" secretes hormone "AB" and autocrine hormone "B" simultaneously. Then endocrine gland "B" feeds back hormones "AB" and "B" to endocrine gland "A".

The principle of hormone diffusion and reaction in endocrine system is provided with the following characteristics:

(1) Information transmission functions. There is no centralized control part, but only one humoral environment for hormone transmission. Moreover, the behaviors of hormone secretion in endocrine glands are only related to the hormone information in the humoral environment. Thus, it is not required a point-to-point communication.

(2) Specificity. A hormone is a kind of chemical component released by endocrine glands. The combination of the chemical component and hormone receptors is specific, i.e., a hormone receptor only reacts to a specific chemical component (hormone). The specificity of the hormone is determined by genetic information.

(3) Synergy and antagonism between hormones. In the endocrine system, endocrine glands release hormones through stimulus in the humoral environment, where synergy and antagonism between hormones emerge from the different effects of hormones on the same gland. These kinds of interactions have an effect to stabilize the system environment.

By employing the features of hormone transmission and interaction, the endocrine system of the human body can achieve rapid adjustment and maintain high adaptability.

3.2.3 On-line Scheduling Model Inspired by the Principle of Hormone Diffusion and Reaction

To solve the on-line scheduling problem, an on-line scheduling model inspired by the principle of hormone diffusion and reaction is proposed (Fig. 3.3). Acting as endocrine glands, the machine BIMCs (called "machine" in the following) play the role of initiators providing optimization and coordination services to AGV BIMCs (called "AGV" in the following), and AGVs play the role of participants providing response and optimization services. In our model, it is assumed that machines and AGVs have the capacity to handle calculation, comparison, and decision-making operations, since the computation power and the optimization capability are embedded in them. In the proposed model, the tasks act as the stimulus in the system; the shop floor environment acts as the humoral environment. In this way, the diffusion and reaction information are similar to the secretion and feedback of hormones.

As illustrated in Fig. 3.3, when operation task "a" is executed on machine "A", the subsequent need for transportation task "a" emerges. Stimulated by that need, machine "A" evaluates the degree of demand for AGVs; then it packages the demand and transportation task information into a hormone; and finally diffuses the hormone into the shop floor environment. When the AGVs sense the hormone information, they are stimulated by the demand and the transportation task simultaneously. In response to the stimulus, each AGV evaluates how the demand influences itself, and its performance to complete the transportation task; then it packages the evaluated value into relevant hormone information and releases this information to machine "A". According to the feedback hormone information, machine "A" allocates the transportation task to the relevant AGV. In the process of on-line scheduling, the

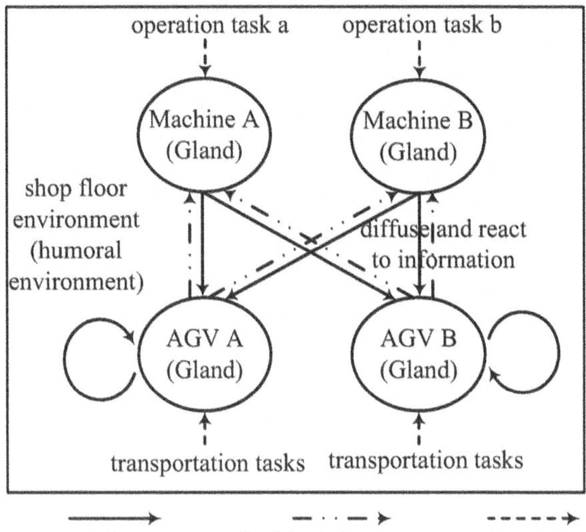

Fig. 3.3 On-line scheduling model inspired by the principle of hormone diffusion and reaction

information evaluation of machines and AGVs is similar to the process of responding to stimulus and secreting hormone, which is a kind of self-adjustment and adaptation mechanism to maintain the balance of the system. By utilizing this model, the transportation tasks can be rapidly allocated without a supervisor. In the next section, the allocation process of transportation tasks is described in detail.

3.3 Allocation Mechanism Based on Hormone Regulation Mechanism

3.3.1 Hormone Regulation Mechanism Background

Farhy [39] defined the basic law for modeling the secretion of hormones by endocrine glands. In this model, the hormone regulation complies with *Hill* function which is composed of the rising function $F_{up}(C)$ and decreasing function $F_{down}(C)$, shown in Eq. (3.1).

$$F(C) = \begin{cases} F_{up}(C) = \dfrac{C^n}{T^n + C^n}, & \text{rising function} \\ F_{down}(C) = \dfrac{T^n}{T^n + C^n}, & \text{decreasing function} \end{cases} \tag{3.1}$$

where C is variable of hormone concentration; T is a threshold of hormone concentration and $T > 0$; n is the *Hill* coefficient and $n > 1$.

If a hormone x is controlled by a set of hormones $I = \{1, 2, \ldots, i\}$ simultaneously, the secretion speed of hormone x is determined by the concentration of these hormones, which is shown in Eq. (3.2).

$$S_x = S_{x0} + \sum_{i=1}^{I} a_i \cdot F(C_i) \tag{3.2}$$

where S_{x0} represents the initial secretion speed of hormone x, C_i is the concentration of hormone i, and coefficient a_i is a positive constant associated with hormone i.

For example, the glucagon secreted by pancreatic islet α-cells has the function of raising blood sugar concentration; on the contrary, the insulin secreted by pancreatic islet β-cells has the function of reducing blood sugar concentration. When the blood sugar concentration is low, islet α-cells promote the glucagon secretion, and hormone regulation complies with up-*Hill* function for rapidly improving the blood sugar concentration; on the contrary, when the blood sugar concentration is high, islet β-cells promote the insulin secretion, and hormone regulation complies with down-*Hill* function for agilely reducing the blood sugar concentration. Because hormone regulation has the characteristics of monotonicity and non-negativity, the affection of one hormone to another hormone can only be the stimulation or inhabitation. In biology, insulin and glucagon exist at the same time, and have the function of mutual antagonism to regulate blood sugar concentration.

If S_x represents the secretion speed of blood sugar, and C_y and C_z are the hormone concentration of insulin and glucagon, respectively, S_x can be calculated according to the influence on blood sugar concentration caused by antagonism between glucagon and insulin.

$$S_x = S_{x0} + a_y F_{\text{up}}(C_y) + a_z F_{\text{down}}(C_z) \tag{3.3}$$

There are not only antagonistic effects but also synergistic effects between different hormones. For example, glucagon and epinephrine have the synergistic function of elevating blood sugar concentration. If S_x represents the secretion speed of blood sugar, and C_y and C_z are the hormone concentration of glucagon and epinephrine, respectively, S_x can be calculated in accordance with the influence on blood sugar concentration caused by synergy between glucagon and epinephrine.

$$S_x = S_{x0} + a_y F_{\text{up}}(C_y) + a_z F_{\text{up}}(C_z) \tag{3.4}$$

The next sections describe the design of the AGV allocation approach inspired by the hormone regulation mechanism.

3.3.2   Time Parameters in Scheduling

In the on-line scheduling problem of machines and AGVs, a task has two main stages: the task being operated by a machine (TO) and the task being transported by an AGV (TT), shown in Fig. 3.4. The notations are given below:

st	is the start time of an operation task.
ct	is the completion time of an operation task.
ot	is the duration of an operation task.
lt	is the loading time of a transportation task.
ult	is the unloading time of a transportation task.
tt	is the duration of a transportation task.
M	is the set of machines.
K	is the set of AGVs.
J	is the set of jobs.

Fig. 3.4 Two main stages of a task

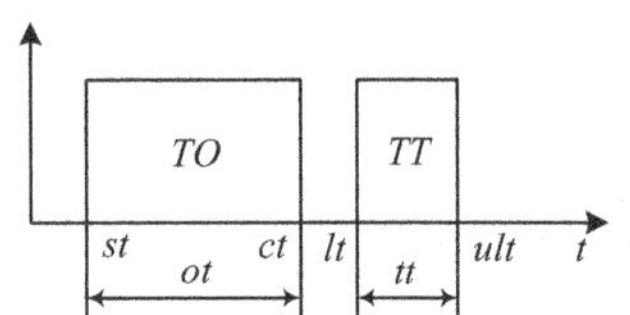

L is the set of operations of a job.
Z is the makespan.

The objective of on-line scheduling problem of machines and AGVs is to minimize Z. According to the interrelation between time parameters, constraints are extracted and formulized as follows:

$$st_{jl+1} \geq ct_{jl} + tt_{jl}, j \in J, l \in L \tag{3.5}$$

$$ct_{jl} + tt_{jl} \leq ult_{jl} \leq st_{jl+1}, j \in J, l \in L \tag{3.6}$$

Formula (3.5) presents a time constraint between adjacent operations of a job; and Formula (3.6) presents time constraints between AGV and adjacent operations of a job. In the on-line scheduling problem, only ot and tt are considered as given parameters, and other parameters are variables to be determined. According to the constraints (3.5) and (3.6), the interrelation between parameters is determined. Taking the operation l of job j as an example, calculation processes of the above parameters are illustrated as follows.

When an operation task (TO_{jl}) is executed on a machine, st_{jl} and ot_{jl} are determined. And then ct_{jl} can be obtained in accordance with Eq. (3.7).

$$ct_{jl} = st_{jl} + ot_{jl} \tag{3.7}$$

According to the operation state of AGV (i.e., transporting a task; or being idle and waiting for a task to transport), the loading time of AGV is described by the following equations:

$$lt_k^{jl} = \begin{cases} \begin{cases} eft_k^{jl} + \Delta t(NL, PL_{jl}), \ eft_k > ept_{jl} \\ eft_k^{jl} + \max\{\Delta t(NL, PL_{jl}), (ept_{jl} - eft_k)\}, \quad eft_k \leq ept_{jl} \end{cases} & \text{transporting state} \\ \begin{cases} t + \Delta t(CL, PL_{jl}), \ t > ept_{jl} \\ t + \max\{\Delta t(CL, PL_{jl}), (ept_{jl} - t)\}, \quad t \leq ept_{jl} \end{cases} & \text{idle or waiting state} \end{cases} \tag{3.8}$$

where t is the current time; eft_k is the earliest finish time of the current transportation task of AGV_k; ept_{jl} is the earliest pickup time of the operation l of job j; NL and CL denote the next location and current location of AGV_k respectively; PL_{jl} is the pickup location of the operation l of job j; and Δt is the required transportation time between two locations. In this paper, the post-processes after TO_{jl} are not taken into account, therefore, the value of ept_{jl} is equal to ct_{jl} ($ept_{jl} = ct_{jl}$). The upper part of Eq. (3.8) defines the situation in which the AGV is transporting a task; and the lower part of Eq. (3.8) defines the situation in which the AGV is idle and waiting to transport a new task.

When the workpiece is picked up by AGV_k, tt_{jl} is calculated as follows:

$$tt_{jl} = \Delta t(PL_{jl}, DL_{jl}) \tag{3.9}$$

where DL_{jl} is the destination location of the operation l of job j.

The unloading time of AGV_k is computed according to Eq. (3.10).

$$ult_{jl} = lt_k^{jl} + tt_{jl} \tag{3.10}$$

Finally, the makespan is selected in accordance with Eq. (3.11)

$$Z = \underset{j \in J}{\mathrm{MAX}}\{ult_{jL_j}\} \tag{3.11}$$

where L_j is the number of the last operation of job j.

3.3.3   Allocation Mechanism

In the case of normal operation, hormone concentrations of the multi-AGV system are in the equilibrium state. Once a transportation task emerges in the manufacturing system, the balance will be disturbed. It is assumed that the secretion of the hormone is triggered in the manufacturing environment, stimulated by the transportation task. The level of hormone concentration is related to the effect on the system balance caused by the initiation of the transportation task. If the effect on the system balance is small, the level of hormone concentration released by the system is low; otherwise, the level of hormone concentration is high. Taking this principle as a basis, a transportation task allocation mechanism is proposed. In the following subsections, the operation l of job j is taken as an example to describe the allocation mechanism in detail.

3.3.3.1   Hormone Secretion of Machines

When an operation task (TO_{jl}) is executed on one machine (M_m), the allocation of transportation task (TT_{jl}) will be triggered; at the same time, M_m will search for the appropriate AGV to perform TT_{jl}. In order to ensure that TT_{jl} can be started after ct_{jl}, which means the workpiece of TT_{jl} is picked up by an AGV at time ct_{jl}, TT_{jl} stimulates M_m to secrete hormone according to Eq. (3.12). The level of hormone concentration represents the urgency of the machine's demand for an AGV.

$$H_m^{jl}(t) = ce^{(t - ct_{jl})/ot_{jl}} \tag{3.12}$$

where c is a positive constant.

An example of hormone secretion curve of a machine is shown in Fig. 3.5 according to Eq. (3.12). When $t < ct_{jl}$, $H_m^{jl}(t) < c$ can be obtained, and the hormone concentration increases moderately as time goes on. It indicates that there is less demand for an AGV to perform TT_{jl}. When $t \geq ct_{jl}$, $H_m^{jl}(t) \geq c$ can also be obtained, and

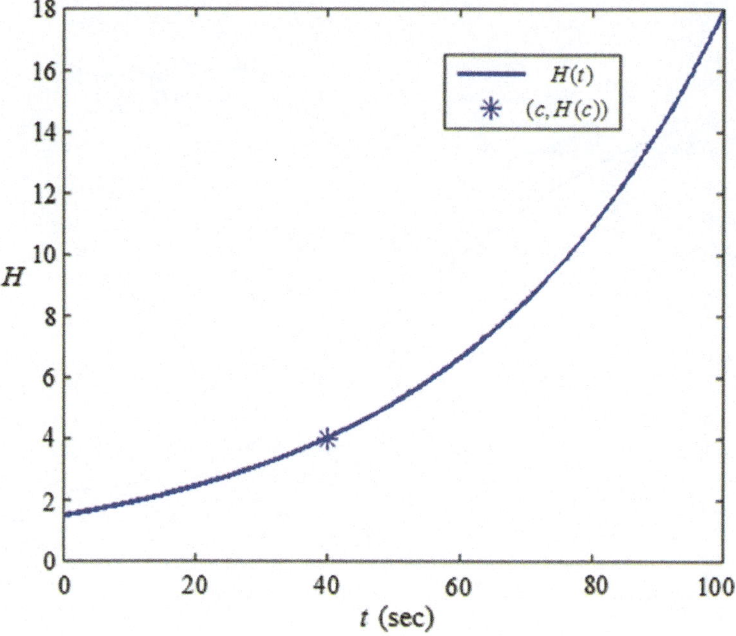

Fig. 3.5 Hormone secretion curve of machine stimulated by transportation task

the hormone concentration increases faster with the passage of time. It demonstrates that the demand for an AGV to perform TT_{jl} is urgent.

3.3.3.2 Hormone Secretion of AGVs

When an AGV with transport capability senses the stimulation of $H_m^{jl}(t)$, the AGV will secrete hormone according to the secretion speed proposed in Eq. (3.13).

$$S_k(H_m^{jl}(t)) = \begin{cases} S_k^{\text{basal}} + a_H F_{\text{down}}(H_m^{jl}(t)), & H_m^{jl}(t) < c \\ S_k^{\text{basal}} + a_H F'_{\text{down}}(H_m^{jl}(t)), & H_m^{jl}(t) \geq c \end{cases} \tag{3.13}$$

where S_k^{basal} is the basal secretion speed of AGV_k and $S_k^{\text{basal}} \geq 0$.

An example of hormone secretion speed curve of AGV stimulated by $H_m^{jl}(t)$ (Fig. 3.6) is plotted in accordance with Eq. (3.13). Affected by the inhibition of $H_m^{jl}(t)$, the secretion speed of AGV_k reduces as time goes on. When $t \geq ct_{jl}$, $H_m^{jl}(t)$ will speed up to inhibit the hormone secretion of AGV_k, to assure that the machine with higher hormone concentration $\left(H_m^{jl}(t)\right)$ can be selected more easily.

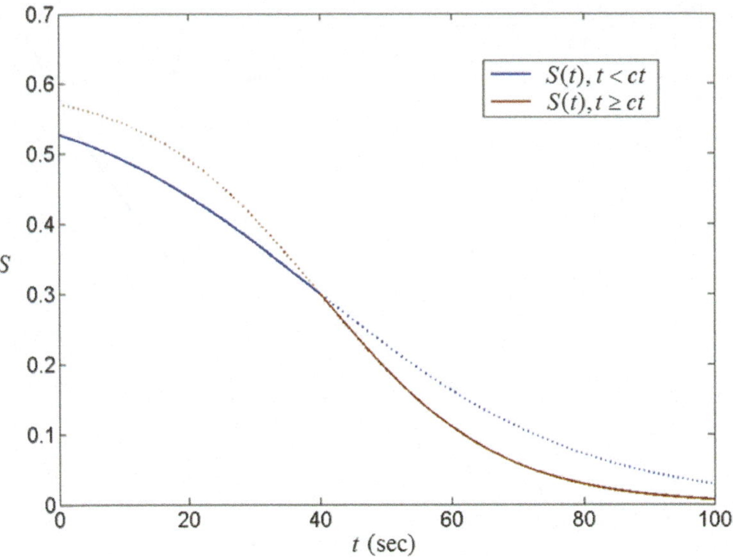

Fig. 3.6 Hormone secretion speed curve of AGV stimulated by $H_m^{jl}(t)$

Besides $H_m^{jl}(t)$, TT_{jl} can also stimulate AGV_k to secrete hormone. In such case, stimulated by TT_{jl}, AGV_k computes the time node (ult_{jl}) to complete TT_{jl} according to Eq. (3.10). The performance of AGV_k to finish TT_{jl} depends on the time interval between ult_{jl} and st'_{jl+1}, where st'_{jl+1} is the earliest and estimated start time of the next operation of job j under the constraints. Thus, stimulated by the performance of an AGV to finish a transportation task, the AGV secretes hormone in accordance with Eq. (3.2) as follows:

$$S_k(ult_{jl}) = \begin{cases} S_k^{\text{basal}} + a_{ult}F_{\text{down}}(ult_{jl}), & ult_{jl} < st'_{jl+1} \\ S_k^{\text{basal}} + a_{ult}F_{\text{up}}(ult_{jl}), & ult_{jl} \geq st'_{jl+1} \end{cases} \tag{3.14}$$

When $ult_{jl} < st'_{jl+1}$, the hormone secretion of AGV_k complies with the down-*Hill* function, otherwise complies with the up-*Hill* function.

An example of hormone secretion speed curve of AGV stimulated by ult (Fig. 3.7) is plotted according to Eq. (3.14). When ult_{jl} is close to st'_{jl+1}, the secretion speed is slower. It indicates that AGV_k can be a better option to complete TT_{jl}, thus it is easier for AGV_k to be selected; on the contrary, when ult_{jl} is far from st'_{jl+1}, the secretion speed is faster. It indicates that AGV_k can be a worse option to complete TT_{jl}, thus it is more difficult for AGV_k to be selected.

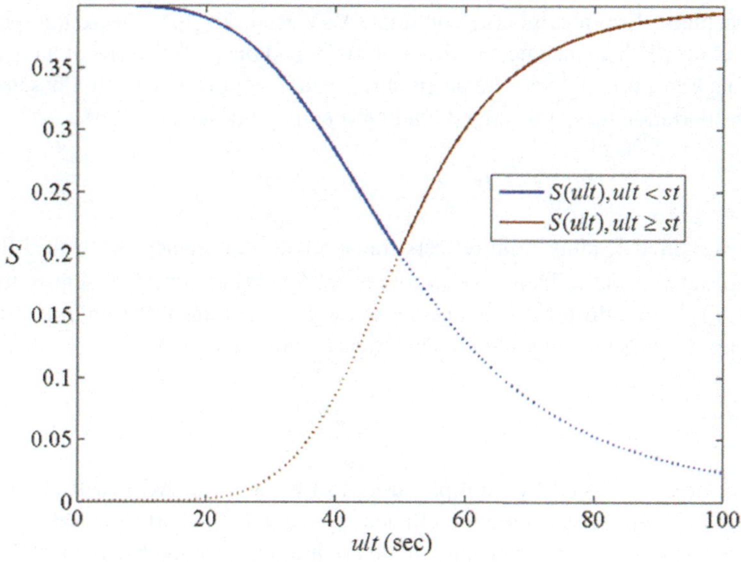

Fig. 3.7   Hormone secretion speed curve of AGV stimulated by *ult*

As mentioned above, the hormone secretion of AGV_k is simultaneously influenced by two independent hormones ($H_m^{jl}(t)$ and ult_{jl}). Thus stimulated by both $H_m^{jl}(t)$ and ult_{jl} at the same time, AGV_k secretes hormone in accordance with the speed following Eq. (3.15).

$$S_k(H_m^{jl}(t), ult_{jl}) = \begin{cases} a_H F_{\text{down}}(H_m^{jl}(t)) + a_{ult} F_{\text{down}}(ult_{jl}) \\ \qquad\qquad + S_k^{basal}, t < ct_{jl}, ult_{jl} < st'_{jl+1} \\ a_H F_{\text{down}}(H_m^{jl}(t)) + a_{ult} F_{\text{up}}(ult_{jl}) \\ \qquad\qquad + S_k^{basal}, t < ct_{jl}, ult_{jl} \geq st'_{jl+1} \\ a_H F'_{\text{down}}(H_m^{jl}(t)) + a_{ult} F_{\text{down}}(ult_{jl}) \\ \qquad\qquad + S_k^{basal}, t \geq ct_{jl}, ult_{jl} < st'_{jl+1} \\ a_H F'_{\text{down}}(H_m^{jl}(t)) + a_{ult} F_{\text{up}}(ult_{jl}) \\ \qquad\qquad + S_k^{basal}, t \geq ct_{jl}, ult_{jl} \geq st'_{jl+1} \end{cases} \tag{3.15}$$

3.3.3.3   Decision-Making

When a machine senses the set $\{S_k(H_m^{jl}(t), ult_{jl})\}$, it will allocate TT_{jl} to an AGV with the minimum secretion speed of hormone at time t according to Eq. (3.16).

$$S_k(t) = \min_{k \in K}\{S_k(H_m^{jl}(t), ult_{jl})\} \tag{3.16}$$

A machine selects the right (i.e., optimal) AGV according to the secretion speed of hormone, in which a machine's need for an AGV and the performance of an AGV to perform a transportation task are taken into account simultaneously. In the selection of a transportation task, a weighted loading time is proposed as follows:

$$l\tilde{t}_k^{jl} = (lt_k^{jl} - t)/H_m^{jl}(t) \tag{3.17}$$

The weighted loading time reflects transportation efficiency between different tasks considered the influence of hormone $H_m^{jl}(t)$. When an AGV senses the set $\{H_m^{jl}(t), TT_{jl}\}$ from different machines, it will select the transportation task with the minimum weighted loading time in accordance with Eq. (3.18).

$$lt_k' = \min_{j \in J, l \in L} \{l\tilde{t}_k^{jl}\} \tag{3.18}$$

Therefore, stimulated by multiple tasks to transport, an AGV will choose the task with the highest transportation efficiency to secrete the hormone. Taking transportation tasks as the communication media of mutual selection between AGVs and machines, the arrival of tasks on time and the transportation efficiency of the AGV system can be ensured simultaneously. In the processes of optimization, AGV can select the next task to transport ahead of schedule, so the coherence of processing between previous and the next operations can be guaranteed.

3.4 Distributed Cooperation Mechanism for On-line Scheduling

In this section, a distributed cooperation mechanism for on-line scheduling is elaborated. This mechanism is mainly to solve mutual selection and optimization between machines and AGVs. The goal of the mechanism is that the distributed system can achieve an optimized production status and maintain a high production level. As shown in Fig. 3.8, machines and AGVs are considered autonomous and cooperative

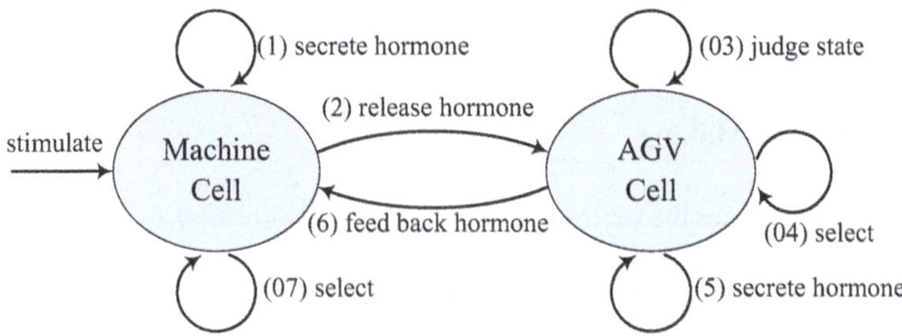

Fig. 3.8 Distributed cooperation mechanism between machines and AGVs

cells [40] that follow the steps given below to conduct their inherent operations:

Step (1) When an operation task $\left(TO_{jl}\right)$ is performed on one machine (M_m) at current time (t), the subsequent transportation task $\left(TT_{jl}\right)$ will be allocated to an AGV by machine cell m. Stimulated by TT_{jl}, machine cell m secretes hormone ($H_m^{jl}(t)$) according to the deviation between the completion time (ct_{jl}) and t.

Step (2) Machine cell m releases hormone information ($H_m^{jl}(t)$, TT_{jl}) into the shop floor environment.

Step (3) When AGV cell k senses the set $\{H_m^{jl}(t), TT_{jl}\}$, it will check its transportation capacity. If the number of transportation tasks is "0" or "1", it means that AGV_k is available for transportation; otherwise, AGV cell k feeds back the maximum hormones to the related machine cells, and goes to Step (6).

Step (4) AGV cell k extracts the required information from the set $\{H_m^{jl}(t), TT_{jl}\}$, and computes the weighted loading time set $\{\tilde{lt}_k^{jl}\}$ according to Eq. (3.17). Then, AGV cell k selects the transportation task with the highest efficiency (TT') in accordance with Eq. (3.18).

Step (5) Stimulated by ($H_m^{jl}(t)$, TT'), AGV cell k calculates the hormone secretion speed ($S_k^{TT'}(t)$) according to Eq. (3.15).

Step (6) AGV cell k feeds back hormone information ($S_k^{TT'}(t)$) to the corresponding machine cell.

Step (7) When machine cell m senses the set $\{S_k(t)\}$ from different AGV cells, it awards TT_{jl} to the AGV with the minimum hormone secretion speed in accordance with Eq. (3.16).

Step (8) A loop between Step (1) to Step (7) is repeated until all the transportation tasks are allocated, then the makespan is selected according to Eq. (3.11).

Using this allocation mechanism, the machines can allocate transportation tasks according to the secretion speed of the hormone, and the AGVs can select transportation tasks with the premise of transportation efficiency. Finally, the objective of distributed and on-line scheduling of machines and AGVs can be achieved.

3.5 Experimental Study

In order to simulate the proposed hormone regulation based approach for on-line scheduling, a series of instances were selected from the benchmarks of Bilge and Ulusoy [13]. These instances were used in both off-line and on-line scheduling by many researchers [11, 13–17, 30]. In this paper, the performance comparison between our proposal against off-line [16] and on-line [30] approaches of the specialized literature was carried out. In the experiments, the performance measure selected was makespan. The performance of the off-line approach is considered as the lower bound of on-line approaches. The constraints and assumptions of the experiments were in the following:

(1) One operation cannot be started before its preceding operations are finished and each machine is provided with sufficient input/output buffer space.
(2) The processing parameters of the operation tasks are deterministic such as operation time and setup time.
(3) All AGVs work with the same speed, and transport a single task at a time.
(4) Preemption is not allowed in transportation, i.e., once task transportation is started, it must be finished without interruption.
(5) Disturbances such as conflict, deadlock, delay, resource malfunction, etc., are not considered.

Let's assume an experimental scenario in which 2 AGVs travel between 4 machines and 1 load/unload (L/U) station. Figure 3.9 shows the layout schemes of the instances; Table 3.1 lists the transportation time matrix of AGV; and Table 3.2 gives the data sets of jobs. In the design of this test problem, the ratio between transportation time and operation time is taken into consideration as a significant characteristic. The ratio can be described as $\overline{tt}/\overline{pt}$, where $\overline{tt}$ represents the average value of the transportation time in the matrix, and $\overline{pt}$ indicates the average value of all the operation time in the job sets. According to the ratio $\left(\overline{tt}/\overline{pt}\right)$, the problems can be divided into two groups: the one with relatively low ratio $\left(\overline{tt}/\overline{pt} < 0.25\right)$ and the other with relatively high ratio $\left(\overline{tt}/\overline{pt} > 0.25\right)$.

All procedures were implemented in JAVA under JDK1.7.0 and the experimental tests were carried out on a 2.0 GHz Intel Core Duo computer with 2 GB Ram under Windows 7. Combining the 10 job sets and the 4 layout schemes from Tables 3.1

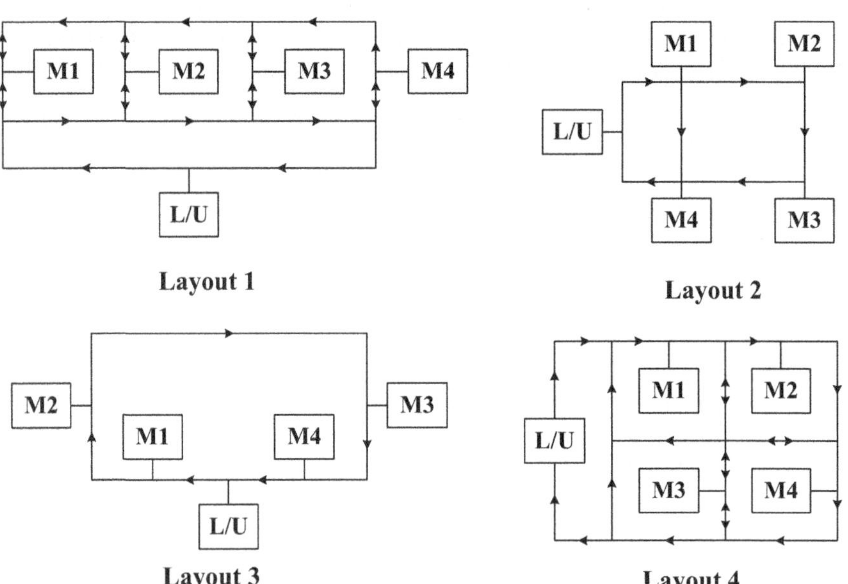

Fig. 3.9 Layout schemes [13]

Table 3.1 Transportation time matrix of AGV [13]

	L/U	M1	M2	M3	M4
Layout 1					
L/U	0	6	8	10	12
M1	12	0	6	8	10
M2	10	6	0	6	8
M3	8	8	6	0	6
M4	6	10	8	6	0
Layout 2					
L/U	0	4	6	8	6
M1	6	0	2	4	2
M2	8	12	0	2	4
M3	6	10	12	0	2
M4	4	8	10	12	0
Layout 3					
L/U	0	2	4	10	12
M1	12	0	2	8	10
M2	10	12	0	6	8
M3	4	6	8	0	2
M4	2	4	6	12	0
Layout 4					
L/U	0	4	8	10	14
M1	18	0	4	6	10
M2	20	14	0	8	6
M3	12	8	6	0	6
M4	14	14	12	6	0

and 3.2, 82 test instances were created. These instances were tested using an off-line heuristic approach (HA) [16], a multi-agent system approach (MAS) [30] and the proposed hormone regulation based approach (HRA). The off-line heuristic approach (HA) determined a lower bound of the makespan, so the aim of the experimental study is to compare our proposal (HRA) against MAS and determine the deviation of both MAS and HRA with respect HA (Dev1, Dev2). Thus, we also show the deviation of HRA with respect to MAS (Dev3). The results for the case $\overline{tt}/\overline{pt} > 0.25$ are listed in Table 3.3, and for the case $\overline{tt}/\overline{pt} < 0.25$ are shown in Table 3.4. The two numbers following EX represent the series of job set and the layout scheme. In Table 3.4, a number "0" or "1" in the last position indicates that the operation times are doubled or tripled, respectively, and in both cases transportation times are halved.

The results show that our approach HRA outperforms MAS in most instances of Tables 3.3 and 3.4. It can be observed that HA obtained the best results in almost

Table 3.2 Data sets of jobs [13]

JobSet1	JobSet2
Job1: M1(8); M2(16); M4(12);	Job1: M1(10); M4(18);
Job2: M1(20); M3(10); M2(18)	Job2: M2(10); M4(18)
Job3: M3(12); M4(8); M1(15)	Job3: M1(10); M3(20)
Job4: M4(14); M2(18)	Job4: M2(10); M3(15); M4(12)
Job5: M3(10); M1(15)	Job5: M1(10); M2(15); M4(12)
	Job6: M1(10); M2(15); M3(12)
JobSet3	JobSet4
Job1: M1(16); M3(15)	Job 1: M4(11); M1(10); M2(7)
Job2: M2(18); M4(15)	Job2: M3(12); M2(10); M4(8)
Job3: M1(20); M2(10)	Job3: M2(7); M3(10); M1(9); M3(8)
Job4: M3(15); M4(10)	Job4: M2(7); M4(8); M1(12); M2(6)
Job5: M1(8); M2(10); M3(15); M4(17)	Job5: M1(9); M2(7); M4(8); M2(10); M3(8)
Job6: M2(10); M3(15); M4(8); M1(15)	
JobSet5	JobSet6
Job1: M1(6); M2(16); M4(9)	Job1: M1(9); M2(11); M4(7)
Job2: M1(18); M3(6); M2(15)	Job2: M1(19); M2(20); M4(13)
Job3: M3(9); M4(3); M1(12)	Job3: M2(14); M3(20); M4(9)
Job4: M4(6); M2(15)	Job4: M2(14); M3(20); M4(9)
Job5: M3(3); M1(9)	Job5: M1(11); M3(16); M4(8)
	Job6: M1(10); M3(12); M4(10)
JobSet7	JobSet8
Job1: M1(6); M3(6)	Job1: M2(12); M3(21); M4(11)
Job2: M2(11); M4(9)	Job2: M2(12); M3(21); M4(11)

(continued)

Table 3.2 (continued)

JobSet1	*JobSet2*
Job3: M2(9); M4(7)	Job3: M2(12); M3(21); M4(11)
Job4: M3(16); M4(7)	Job4: M2(12); M3(21); M4(11)
Job5: M1(9); M3(18)	Job5: M1(10); M2(14); M3(18); M4(9)
Job6: M2(13); M3(19); M4(6)	Job6: M1(10); M2(14); M3(18); M4(9)
Job7: M1(10); M2(9); M3(13)	
Job8: M1(11); M2(9); M4(8)	
JobSet9	*JobSet10*
Job1: M3(9); M1(12); M2(9); M4(6)	Job1: M1(11); M3(19); M2(16); M4(13)
Job2: M3(16); M2(11); M4(9)	Job2: M2(21); M3(16); M4(14)
Job3: M1(21); M2(18); M4(7)	Job3: M3(8); M2(10); M1(14); M4(9)
Job4: M2(20); M3(22); M4(11)	Job4: M2(13); M3(20); M4(10)
Job5: M3(14); M1(16); M2(13); M4(9)	Job5: M1(9); M3(16); M4(18)
	Job6: M2(19); M1(21); M3(11); M4(15)

all instances due to the fact that it is an off-line heuristic and all information on the scheduling problem is known in advance.

From the results in Table 3.3, the mean total deviation (MTD) of HRA to HA is 33.83%, which means that HRA does not go beyond HA in makespan. This result is in accordance with our expectations as HRA is not an off-line algorithm but on-line and distributed approach. Compared with an on-line approach (MAS), 55% of the results from HRA are better than MAS, and the MTD of HRA to MAS is -1.65%. Thus, HRA shows better performance in makespan. From the results in Table 3.4, the performance of HRA is better. Compared with on-line approach, the performance of HRA is better than MAS in 69% test instances, and the MTD of HRA to MAS is -5.28%. Because the MTD of HRA to HA is 5.23%, it shows that the optimization capability of our approach is closed to off-line method. Similar results are obtained for EX110, EX220, EX630, etc. In other instances of EX820, it shows the same performance with HA. Similar results are obtained for EX 620, EX830, and EX 840. By comparing the results, it is clear that HRA exhibits better performance in the case of a relatively low $\overline{tt}/\overline{pt}$ ratio. As discussed above, the HRA shows the potential to improve the performance in terms of distributed on-line scheduling of machines and AGVs.

Table 3.3 Comparison results in the case of $\overline{tt}\,/\,\overline{pt} > 0.25$

Instance	$\overline{tt}\,/\,\overline{pt}$	HA [16]	MAS [30]		HRA		
		Makespan	Makespan	Dev1%	Makespan	Dev2%	Dev3%
EX11	0.59	**96**	130	35.42	122	27.08	−6.15
EX21	0.61	**100**	143	43.00	140	40.00	−2.10
EX31	0.59	**99**	142	43.43	148	49.49	4.23
EX41	0.91	**112**	198	76.79	174	55.36	−12.12
EX51	0.85	**87**	130	49.43	129	48.26	−0.77
EX61	0.78	**118**	153	29.66	153	29.66	0.00
EX71	0.78	**111**	129	16.22	146	31.53	13.18
EX81	0.58	**161**	196	21.74	199	23.60	1.53
EX91	0.61	**116**	178	53.45	147	26.72	−17.42
EX101	0.55	**147**	188	27.89	175	19.05	−6.91
EX12	0.47	**82**	98	19.51	100	21.95	2.04
EX22	0.49	**76**	86	13.16	86	13.16	0.00
EX32	0.47	**85**	114	34.12	114	34.12	0.00
EX42	0.73	**87**	129	48.28	120	37.93	−6.98
EX52	0.68	**69**	98	42.03	100	44.93	2.04
EX62	0.54	**98**	123	25.51	106	8.16	−13.82
EX72	0.62	**79**	92	16.46	101	27.85	9.78
EX82	0.46	**151**	172	13.91	156	3.31	−9.30
EX92	0.49	**102**	123	20.59	127	24.51	3.25
EX102	0.44	**135**	154	14.07	158	17.04	2.60
EX13	0.52	**84**	109	29.76	102	21.43	−6.42
EX23	0.54	**86**	98	13.95	96	11.63	−2.04
EX33	0.51	**86**	103	19.77	115	33.72	11.65
EX43	0.80	**89**	155	74.16	127	42.70	−18.06
EX53	0.74	**74**	109	47.30	113	52.70	3.67
EX63	0.54	**103**	128	24.27	110	6.80	−14.06
EX73	0.68	**83**	93	12.05	119	43.37	27.96
EX83	0.50	**153**	172	12.42	167	9.15	−2.91
EX93	0.53	**105**	119	13.33	122	16.19	2.52
EX103	0.49	**139**	158	13.67	159	14.39	0.63
EX14	0.74	**103**	168	63.11	162	57.28	−3.57
EX24	0.77	**108**	169	56.48	156	44.44	−7.69
EX34	0.74	**111**	167	50.45	170	53.15	1.80
EX44	1.14	**126**	242	92.06	217	72.22	−10.33
EX54	1.06	**96**	168	75.00	160	66.67	−4.76

(continued)

Table 3.3 (continued)

Instance	$\overline{tt}/\overline{pt}$	HA [16]	MAS [30]		HRA		
		Makespan	Makespan	Dev1%	Makespan	Dev2%	Dev3%
EX64	0.78	**120**	189	57.50	182	51.67	−3.70
EX74	0.97	**126**	156	23.81	180	42.86	15.38
EX84	0.72	**163**	251	53.99	243	49.08	−3.19
EX94	0.76	**122**	181	48.36	170	39.34	−6.08
EX104	0.69	**158**	246	55.70	222	40.51	−9.76
MTD				**37.04**		**33.83**	**−1.65**

The Gantt chart of instance EX23 is shown in Fig. 3.10. The numbers in the Gantt chart represent the serial number of tasks. For example, the number "21" in M2 represents the first operation task of job2; the number "21" in AGV1 indicates the first transportation of job2. As can be seen from the Gantt chart, at the time 48, a new round of mutual selection between machines and AGVs starts; two AGVs have the transportation capacity to accept more transportation tasks; and operations "61", "52", and "32" demand AGVs for transportation. In the AGV selection stage, affected by the hormone H and loading time, the two AGVs select the operation "61" with the lowest weighted loading time; then in the machine selection stage, AGV2 is selected as the winner. Because compared with AGV1, AGV2 can finish the task with a lower deviation, and the hormone secretion speed is lower than AGV1. The mean value of the transportation task of AGV is 11.2 which can be computed by using the data in Table 3.1. As shown in the Gantt chart, 21 transportation tasks are implemented by two AGVs, and the distribution of time for AGVs is shown in Fig. 3.11. 90.48% of the transportation tasks have time consumption of less than 15. As shown in Fig. 3.12, the utilization rate of two AGVs remains at a high and balanced level.

3.6 Conclusions

In this paper, a hormone regulation based approach for distributed and on-line scheduling of machines and AGVs is proposed. The principle of hormone diffusion and reaction in the endocrine system provides a new information processing method for the on-line scheduling model to agilely cope with new arrival tasks. According to the hormone regulation mechanism, the deviations between completion time and planned time in on-line allocation processes are used to optimize makespan. Thus, in the mutual selection between machines and AGVs, machines can allocate transportation tasks according to the minimum secretion speed of hormone, and AGVs can select transportation tasks with the highest transportation efficiency. As the tasks are well allocated and good performances of the system can be obtained.

Table 3.4 Comparison results in the case of $\overline{tt}/\overline{pt} < 0.25$

Instance	$\overline{tt}/\overline{pt}$	HA [16]	MAS [30]		HRA		
		Makespan	Makespan	Dev1%	Makespan	Dev2%	Dev3%
EX110	0.15	**126**	135	7.14	131	3.97	−2.96
EX210	0.15	**148**	157	6.08	151	2.03	−3.82
EX310	0.15	**150**	154	2.67	156	4.00	1.30
EX410	0.15	**119**	211	77.31	129	8.40	−38.86
EX510	0.21	**102**	118	15.69	108	5.88	−8.47
EX610	0.16	**186**	204	9.68	189	1.61	−7.35
EX710	0.19	**137**	138	0.73	150	9.49	8.70
EX810	0.14	**272**	330	21.32	287	5.51	−13.03
EX910	0.15	**176**	191	8.52	190	7.95	−0.52
EX1010	0.14	**238**	269	13.03	260	9.24	−3.35
EX120	0.12	**123**	127	3.25	127	3.25	0.00
EX220	0.12	**143**	151	5.59	145	1.40	−3.97
EX320	0.12	145	**144**	−0.69	148	2.07	2.78
EX420	0.12	**114**	161	41.23	121	6.14	−24.84
EX520	0.17	**100**	110	10.00	104	4.00	−5.45
EX620	0.12	**181**	196	8.29	**181**	0.00	−7.65
EX720	0.15	136	**132**	−2.94	144	5.88	9.09
EX820	0.11	**287**	319	11.15	**287**	0.00	−10.03
EX920	0.12	**173**	187	8.09	181	4.62	−3.21
EX1020	0.11	**236**	266	12.71	250	5.93	−6.02
EX130	0.13	**122**	134	9.84	127	4.10	−5.22
EX230	0.13	**146**	151	3.42	147	0.68	−2.65
EX330	0.13	146	**129**	−11.64	154	5.48	19.38
EX430	0.13	**114**	228	100.00	126	10.53	−44.74
EX530	0.18	**99**	111	12.12	110	11.11	−0.90
EX630	0.14	**182**	198	8.79	183	0.55	−7.58
EX730	0.17	**137**	132	−3.65	146	6.57	10.61
EX830	0.13	288	**273**	−5.21	288	0.00	5.49
EX930	0.13	**174**	187	7.47	183	5.17	−2.14
EX1030	0.12	**237**	266	12.24	245	3.38	−7.89
EX140	0.18	**124**	137	10.48	138	11.29	0.73
EX241	0.13	**217**	230	5.99	224	3.23	−2.61
EX340	0.18	**151**	155	2.65	167	10.60	7.74
EX341	0.12	**221**	227	2.71	232	4.98	2.20
EX441	0.19	**172**	344	100.00	187	8.72	−45.64

(continued)

Table 3.4 (continued)

Instance	$\overline{tt}/\overline{pt}$	HA [16]	MAS [30]		HRA		
		Makespan	Makespan	Dev1%	Makespan	Dev2%	Dev3%
EX541	0.18	**148**	158	6.76	158	6.76	0.00
EX640	0.19	**184**	211	14.67	189	2.72	−10.43
EX740	0.24	**137**	158	15.337	154	12.41	−2.53
EX741	0.16	**203**	206	1.48	214	5.42	3.88
EX840	0.18	**293**	331	12.97	**293**	0	−11.48
EX940	0.19	**175**	195	11.43	191	9.14	−2.051
EX1040	0.17	**240**	276	15.00	253	5.42	−8.33
MTD				**14.09**		**5.23**	**−5.28**

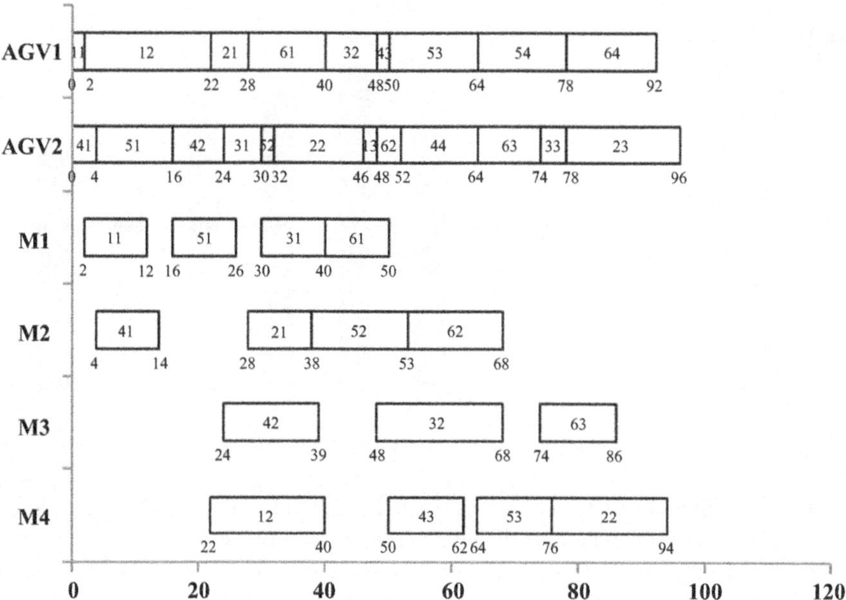

Fig. 3.10 Gantt chart of EX23

The proposed approach was tested and compared with classical instances in the literature. The performance of HRA was close to off-line approaches. Especially in the low ratio, the performance in some instances was the same with off-line approaches. Compared with the on-line approach, the performance of HRA was better in more than half of the test instances, and the rest was the same or close to. The results show that HRA reflects a promising performance in terms of distributed and on-line scheduling of machines and AGVs. The application of hormone regulation in this area shows great potential. Especially its reaction and diffusion mechanisms can

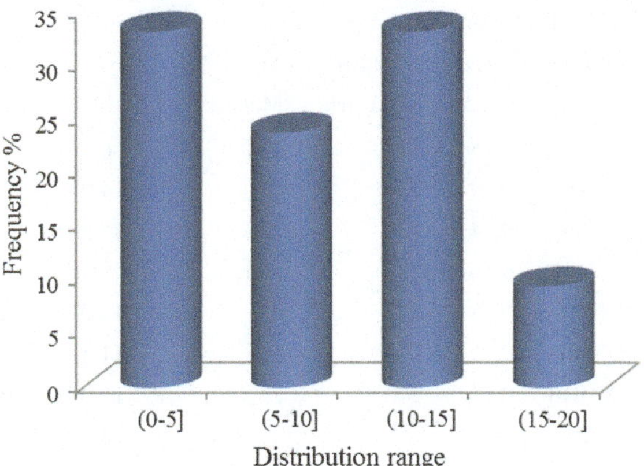

Fig. 3.11 Distribution of time for AGVs

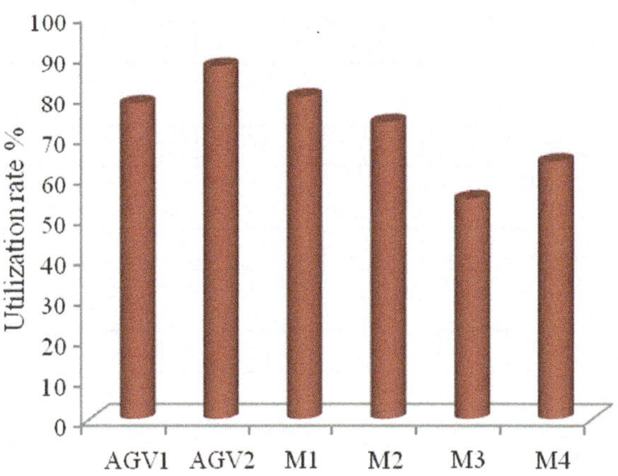

Fig. 3.12 Utilization rates of AGVs and machines

quickly respond and satisfy scheduling requirements in the system. And it expands the view of research on dynamic scheduling problems. Future research can be directed towards the decision-making modules in different stages to improve the performances of the scheduling approach. Besides, the proposed approach can be enhanced to take into account conflict, deadlock, delay, recourse malfunction, rush orders, etc.

References

1. Lin, L., Shinn, S. W., Gen, M., et al. (2006). Network model and effective evolutionary approach for AGV dispatching in manufacturing system. *Journal of Intelligent Manufacturing, 17*(4), 465–477.
2. Qiu, L., Hsu, W. J., Huang, S. Y., et al. (2002). Scheduling and routing algorithms for AGVs: A survey. *International Journal of Production Research, 40*(3), 745–760.
3. Dotoli, M., & Fanti, M. P. (2004). Coloured timed Petri net model for real-time control of automated guided vehicle systems. *International Journal of Production Research, 42*(9), 1787–1814.

4. Mohammed, A., & Khan, W. U. (2010). Implementation issues of AGVs in flexible manufacturing system: A review. *Global Journal of Flexible Systems Management, 11*(1–2), 55–62.
5. Singh, N., Sarngadharan, P. V., & Pal, P. K. (2011). AGV scheduling for automated material distribution: A case study. *Journal of Intelligent Manufacturing, 22*(2), 219–228.
6. Shah, M., Lin, L., & Nagi, R. (1997). A production order-driven AGV control model with object-oriented implementation. *Computer Integrated Manufacturing Systems, 10*(96), 35–48.
7. Blazevicz, J., Eiselt, H. A., Finke, G., et al. (1991). Scheduling tasks and vehicles in a flexible manufacturing system. *International Journal of Flexible Manufacturing System, 4*(1), 5–16.
8. Kesen, S. E., Das, S. K., & Güngör, Z. (2010). A genetic algorithm based heuristic for scheduling of virtual manufacturing cells (VMCs). *Computers & Operations Research, 37*(6), 1148–1156.
9. Vijay Kumar, M., Murthy, A. N. N., & Chandrasekhara, K. (2011). Dynamic scheduling of flexible manufacturing system using heuristic approach. *Opsearch, 48*(1), 1–19.
10. Bülbül, K. (2011). A hybrid shifting bottleneck-tabu search heuristic for the job shop total weighted tardiness problem. *Computers & Operations Research, 38*(6), 967–983.
11. Babu, A. G., Jerald, J., Haq, A. N., et al. (2010). Scheduling of machines and automated guided vehicles in FMS using differential evolution. *International Journal of Production Research, 48*(16), 4683–4699.
12. Confessore, G., Fabiano, M., & Liotta, G. (2013). A network flow based heuristic approach for optimising AGV movements. *Journal of Intelligent Manufacturing, 24*(2), 405–419.
13. Bilge, U., & Ulusoy, G. (1995). A time window approach to simultaneous scheduling of machines and material handling system in an FMS. *Operations Research, 43*(6), 1058–1070.
14. Ulusoy, G., Sivrikaya-Şerifoğlu, F., & Bilge, U. (1997). A genetic algorithm approach to the simultaneous scheduling of machines and automated guided vehicles. *Computers & Operations Research, 24*(4), 335–351.
15. Abdelmaguid, T. F., Nassef, A. O., Kamal, B. A., et al. (2004). A hybrid GA/heuristic approach to the simultaneous scheduling of machines and automated guided vehicles. *International Journal of Production Research, 42*(2), 267–281.
16. Reddy, B. S. P., & Rao, C. S. P. (2006). A hybrid multi-objective GA for simultaneous scheduling of machines and AGVs in FMS. *International Journal of Advanced Manufacturing Technology, 31*(5), 602–613.
17. Lacomme, P., Larabi, M., & Tchernev, N. (2013). Job-shop based framework for simultaneous scheduling of machines and automated guided vehicles. *International Journal of Production Economics, 143*(1), 24–34.
18. Sabuncuoglu, I., & Hommertzheim, D. L. (1992). Dynamic dispatching algorithm for scheduling machines and automated guided vehicles in a flexible manufacturing system. *International Journal of Production Research, 30*(5), 1059–1079.
19. Le-Anh, T., & De Koster, M. B. M. (2006). A review of design and control of automated guided vehicle systems. *European Journal of Operational Research, 171*(1), 1–23.
20. Vis, I. F. A. (2006). Survey of research in the design and control of automated guided vehicle systems. *European Journal of Operational Research, 170*(3), 677–709.
21. Ghasemzadeh, H., Behrangi, E., & Azgomi, M. A. (2009). Conflict-free scheduling and routing of automated guided vehicles in mesh topologies. *Robotics & Autonomous Systems, 57*(6–7), 738–748.
22. Ho, Y., Liu, H., & Yih, Y. (2012). A multiple-attribute method for concurrently solving the pickup-dispatching problem and the load-selection problem of multiple-load AGVs. *Journal of Manufacturing Systems, 31*(3), 288–300.
23. Lee, J., Choi, S., Lee, C., et al. (2001). A study for AGV steering control and identification using vision system. In *Proceedings of IEEE International Symposium on Industrial Electronics*, Pusan, Korea, 12–16 June 2001 (Vol. 3, pp. 1575–1578). IEEE.
24. Lau, Y. K., & Woo, H. S. O. (2007). An agent-based dynamic routing strategy for automated material handling systems. *International Journal of Computer Integrated Manufacturing, 21*(3), 269–288.

25. Morten, L., & Olivier, R. D. (2011). Holonic shop-floor application for handling, feeding and transportation of workpieces. *International Journal of Production Research, 49*(5), 1441–1454.
26. Babiceanu, R. F., Chen, F. F., et al. (2005). Real-time holonic scheduling of material handling operations in a dynamic manufacturing environment. *Robotics and Computer-Integrated Manufacturing, 21*(4–5), 328–337.
27. Singh, P., & Tiwari, M. K. S. (2010). Intelligent agent framework to determine the optimal conflict-free path for an automated guided vehicles system. *International Journal of Production Research, 40*(16), 4195–4223.
28. Breton, L., Maza, S., & Castagna, P. (2006). A multi-agent based conflict-free routing approach of bi-directional automated guided vehicles. In *Proceedings of the 2006 American Control Conference*, Minneapolis, Minnesota, USA, 14–16 June 2006 (pp. 2825–2830). IEEE.
29. Wallace, A. (2007). Multi-agent negotiation strategies utilizing heuristics for the flow of AGVs. *International Journal of Production Research, 45*(2), 309–322.
30. Erol, R., Sahin, C., Baykasoglu, A., et al. (2012). A multi-agent based approach to dynamic scheduling of machines and automated guided vehicles in manufacturing systems. *Applied Soft Computing, 12*(6), 1720–1732.
31. Zahadat, P., & Schmickl, T. (2014). Generation of diversity in a reaction-diffusion-based controller. *Artificial Life, 20*(3), 319–342.
32. Hamann, H., Stradner, J., Schmickl, T., et al. (2010). Artificial hormone reaction networks: Towards higher evolvability in evolutionary multi-modular robotics. In *Proceedings of the ALife XII Conference*, MIT Press, Cambridge, 2010 (pp. 773–780).
33. Zheng, K., Tang, D. B., Giret, A., et al. (2018). A hormone regulation–based approach for distributed and on-line scheduling of machines and automated guided vehicles. *Proceedings of the Institution of Mechanical Engineers, Part B: Journal of Engineering Manufacture, 232*(1), 99–113.
34. Qin, L., & Kan, S. (2013). Production dynamic scheduling method based on improved contract net of multi-agent. In Z. Du (Ed.), *Intelligence computation and evolutionary computation* (pp. 929–936). Berlin, Heidelberg: Springer.
35. Rajabinasab, A., & Mansour, S. (2011). Dynamic flexible job shop scheduling with alternative process plans: An agent-based approach. *International Journal of Advanced Manufacturing Technology, 54*(9), 1091–1107.
36. Kouider, A., & Bouzouia, B. (2012). Multi-agent job shop scheduling system based on cooperative approach of idle time minimisation. *International Journal of Production Research, 50*(2), 409–424.
37. Zheng, K., Tang, D., Giret, A., et al. (2015). Dynamic shop floor re-scheduling approach inspired by a neuroendocrine regulation mechanism. *Proceedings of the Institution of Mechanical Engineers, Part B: Journal of Engineering Manufacture, 229*(S1), 121–134.
38. Budilova, E. V., & Teriokhin, A. T. (1992). Endocrine networks. *IEEE Symposium on Neuroinformatics and Neurocomputers, 2*, 729–737.
39. Farhy, L. S. (2004). Modeling of oscillations of endocrine networks with feedback. *Methods Enzymology, 384*, 54–81.
40. Tang, D., Gu, W., Wang, L., et al. (2011). A neuroendocrine-inspired approach for adaptive manufacturing system control. *International Journal of Production Research, 49*(5), 1255–1268.

Chapter 4
Production Control Strategy Inspired by Neuroendocrine Regulation

4.1 Introduction and Synopsis

Recently, much research on the control strategy of the manufacturing system has been conducted around the world. With the market competition obviously characterized by globalization, dynamism, and customer-driven, manufacturing industry is faced with opportunities brought by technology advancement and worldwide challenge from personalization and diversity tends [1, 2]. Production operating environment is filled with uncertainty owning to shortening product life cycle, smaller production lot, and higher refresh speed. Multi-objective or uncertain incidents in a complex production process are frequently brought up by processing equipment and complicated constraints from process and resource [3]. Unpredictable and dynamic market changes cause changing requirements concerning the output capacity and variety of processing functions of manufacturing systems [4, 5].

Meanwhile, the dynamics and nonlinearity of production load are caused by continuous and unpredictable task, for example, external incidents consisting of production task change or rush order, and internal incidents covering machine malfunction and manufacturing resources interruption [6]. Ideally in an adaptive manufacturing system, exact capacity is supplied when needed and where needed [5]. Thus, the production capacity should be adjusted to meet the demand precisely and continually while keeping it in a profitable state. However, this type of policy is undesirable or impossible due to the fact that the rate of variation in demand is usually much higher than the rate at which capacity can be changed using current production control methods. Current production planning and control (PPC) methods often do not deal with unplanned orders and other types of turbulence in a satisfactory manner [7]. To avoid these kinds of situations, material flow management and capacity scalability are very important and normally used [8].

© Springer Nature Singapore Pte Ltd. 2020
D. Tang et al., *Adaptive Control of Bio-Inspired Manufacturing Systems*,
Research on Intelligent Manufacturing,
https://doi.org/10.1007/978-981-15-3445-4_4

Bio-cybernetics has been increasingly highlighted in recent years, and many kinds of biological intelligent controllers have been presented. Research results in bio-cybernetics prove that the control system in the human body is a perfect system with superior construction and excellent function. As the control center of hormones, the neuroendocrine system is featured with self-adaptation and stability, which exhibits hormone regulation behavior [4, 9]. An extraordinary number of scientific advances have provided a much-improved understanding of the endocrine system, in which individual components communicate with one another in a highly coordinated fashion [10].

In this paper, based on the control advantages of the neuroendocrine system, a new dynamic control model of the production system is used to design and analyze the production control strategy [11]. The aim of this research is to improve the performance of production control, especially the response to disturbances such as rush orders in backlog control of the production system. Motivated by the desire for performance improvement, a backlog controller and a WIP (work-in-progress) controller are designed by utilizing the mechanism of hormone regulation. Then, control-theoretic approaches are used to analyze backlog and WIP controllers of the production system in order to eliminate the backlog quickly and to keep the WIP at a defined level in an adaptive manner.

4.2 Literature Review

Production control systems are characterized by the time-varying nature of the inputs and the response (performance). For dealing with time-varying demand changes (such as rush order) of the manufacturing system, significant past researches have been implemented in the control of capacity and backlog. Towill et al. [12] used classical control concepts to study decision support systems in an adaptive model. Further extensive work was conducted by Qiu et al. [13], who used the linear programming approach to address problems such as capacity, lot size, and cost control to analyze the system performance related to the parameters in a model.

Asl and Ulsory [14] presented a dynamic approach to adjust the capacity to meet the demands specified by marketing based on the use of feedback control. Deif and EIMaraghy [15] proposed an optimal capacity scalability scheduling approach based on the demand and the cost of capacity scalability. Also, Deif and EIMaraghy [5] presented a dynamic model to analyze capacity scalability using transfer function in the reconfigurable manufacturing system, three types of capacity controllers were designed to improve the dynamic performance in response to sudden demand changes. Guo and Jing [16] designed a backlog controller and an input rate controller applying the proportional–integral (PI) control algorithm in discrete manufacturing systems, both controllers interacted with each other to maintain lower and steadier WIP so as to improve on-time delivery.

The funnel model was developed by Wiendahl [17] using control theory to link the external and internal requirements of the manufacturing system with the manipulated variables of a PPC system. An automatic PPC model was presented by Wiendahl and Breithaupt [18, 19] with the help of the funnel model and the theory of the logistic operating curve developed by Nyhuis [20]. In their research, the backlog controller and WIP controller interacted to adjust the capacity and the input rate of the system to eliminate the backlog while maintaining at a planned level of WIP. Simulation studies were used to analyze the effect of a disturbance on system variables such as WIP, capacity, and input rate.

A closed-loop PPC system with autonomous controllers for WIP and capacity was proposed by Duffie and Falu [7]. Transfer function analysis was used to model dynamic relationships between system inputs and variables including backlog and WIP, the results were used to select control laws for desired system performance and to calculate system response. Kim and Duffie [6] extended the previous work to study the impact of capacity disturbances and capacity delays on system performance with the methods of frequency response analysis, a discrete dynamic model was used to design and analyze control algorithms for closed-loop PPC. Their results indicated that if capacity can be adjusted precisely, the manufacturing system's performance would be improved during a time-varying demand condition.

In the previous research works, the production control problems were modeled using conventional control-theoretic approaches (the parameters of the controllers are fixed), which cannot cope with dynamic changes (such as rush orders) in an adaptive manner. Production control has become more challenging as manufacturing systems adapt to frequent demand variation, especially there is a need to model backlog controller to deal with rush orders adaptively and to solve the problem of production oscillation with the WIP controller. To answer the challenges, the production control strategy still deserves extensive research.

It is well known that the neuroendocrine system (NES) can regulate the functions of several organs and glands with high self-adaptability and stability under stimulus, by means of regulating their hormone secretions synchronously [21]. Based on such a mechanism, some recent researches focus on how to use feedback loop and hormone release of NES to design some new control structures and systems. A two-level structure controller based on NES regulation was designed by Liu et al. [22], which can not only achieve accurate control but also adjust control parameters in real time. Extensive work was conducted by Guo et al. [23], an NES-based cooperative intelligent control system was designed to improve the performance of a complex multichannel plant.

Compared with the conventional control system, NES-based control systems have better control performances with hormone regulation, especially under the occurrence of disturbances. In manufacturing systems, an NES-inspired approach for adaptive manufacturing control system was presented by Tang et al. [4]. Following our previous research work [4], this paper presents an adaptive dynamic model for backlog and WIP control from a bio-inspired perspective utilizing the hormone regulation mechanism.

4.3   General Principle of Neuroendocrine System

Neuroendocrinology arose from the recognition that the brain, especially the hypothalamus, controls secretion of pituitary gland hormones, and has subsequently expanded to investigate numerous interconnections of the endocrine and nervous systems. In the humoral regulation process, the human body's physiological parameters maintaining in a stable condition, is mainly on the basis of neuroendocrine hormone regulation mechanism.

4.3.1   Negative Feedback Mechanism of Hormone Regulation

The regulatory process of neuroendocrine hormone is described as follows: through the control of the central nervous system, various endocrine glands (such as thyroid, adrenal glands, and other target organs) through the blood circulation of body fluid circulation system to produce all kinds of hormone regulation and control function, such as insulin adaptive adjustment of blood sugar levels. Meanwhile, the concentration of the hormone is controlled precisely and stably by a negative feedback control loop of hormone regulation.

Metabolic regulation is greatly affected by the adrenal hormone, and the regulation is functioned as follows: corticotrophin-releasing hormone (CRH) is secreted by the hypothalamus, stimulating pituitary release adrenocorticotropic hormone (ACTH), and stimulated by ACTH, the adrenal gland secretes adrenaline. Informed by sensors detecting the improvement of adrenaline concentration, hypothalamus and pituitary reduce the secretion of CRH and ACTH, decrease the concentration of adrenaline and achieve a balance at last. A typical control model of hormone regulation is shown in Fig. 4.1, which exhibits the characteristic of negative feedback.

4.3.2   Hill Functions of Hormone Regulation

Farhy [9] pointed out that the general rule of hormone regulation had characteristics with monotone and nonnegative. A control function with nonlinearity is used, i.e., the up-regulatory or down-regulatory Hill functions, as shown below

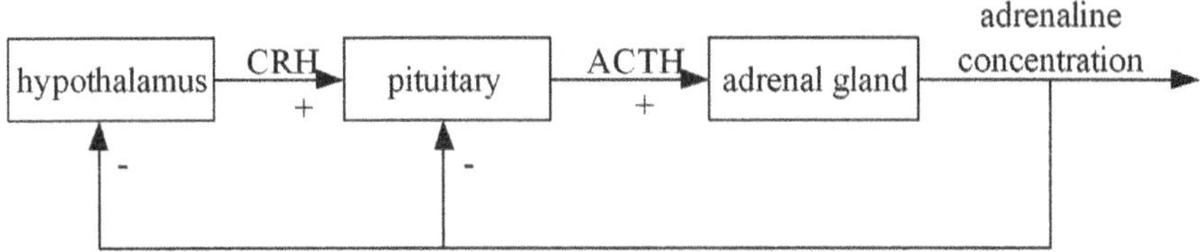

Fig. 4.1 Negative feedback control model of hormone regulation

$$F_{\text{up}}(G) = \frac{G^n}{T^n + G^n} \tag{4.1}$$

$$F_{\text{down}}(G) = \frac{T^n}{T^n + G^n} \tag{4.2}$$

where T is a threshold, $T > 0$, G is an independent variable, n is a Hill coefficient, and $n \geq 1$. Note that $F_{\text{up}} + F_{\text{down}} = 1$ and $0 \leq F_{\text{up}}(G) \leq 1, 0 \leq F_{\text{down}}(G) \leq 1$. The Hill functions can realize quick stability, which keeps hormone regulation adaptive and stable. If one hormone x is controlled by another hormone y, the secretion of the former S_x is determined by the concentration of the latter C_y, which can be described as

$$S_{x(\text{up/down})} = \alpha F(C_y) + S_{x0} \tag{4.3}$$

where S_{x0} is the basal secretion of hormone x, and α is a constant.

4.4 Control Model of Production System

As described in the section above, the hormone regulation mechanism has been adopted to react to disturbances in an emergent state. In this section, a control model of the production system is illustrated with a focus on hormone regulation. The modeling approach and its analysis are based on the application of control theory and hormone regulation mechanism.

4.4.1 Hormone Regulation Model of Production System

Inspired by the hormone regulation mechanism, a control model of the production system is proposed in Fig. 4.2, which is composed of three main components, the production process, WIP controller and backlog controller. Both controllers constitute a kind of two-level coordination architecture of the production system, the backlog controller is in the upper level, and the WIP controller is lower, where the control strategies of backlog and WIP are based on hormone regulation. With the backlog controller and the WIP controller, we assume that physical limits on WIP, input rate, and capacity are not exceeded. The dynamics of the control model can be analyzed with respect to the response to work disturbances. Ideally, the backlog is zero in a balanced production system. The detailed will be introduced in the following subsection.

The input and output variables of the control model of the production system are introduced as follows. The incoming orders of a balanced system are simply converted into the initial input rate and then flow into the backlog controller as one

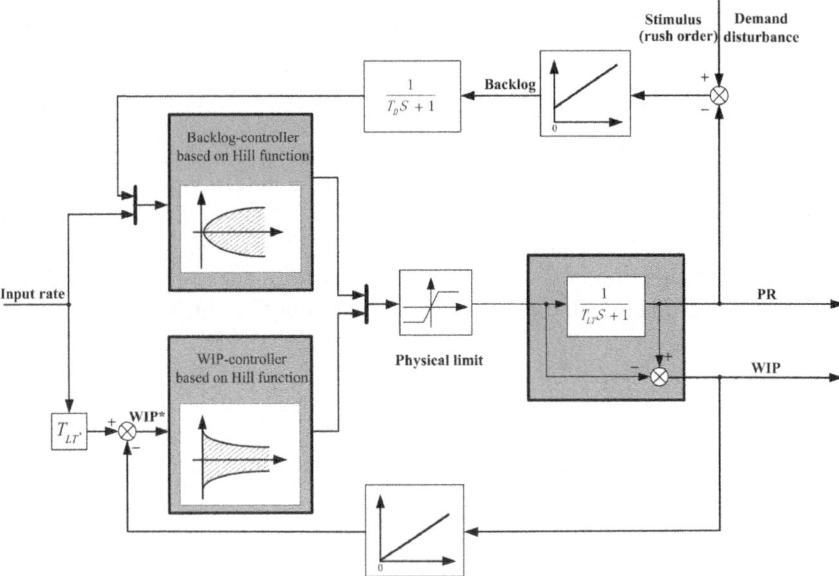

Fig. 4.2 Control model of the production system

of the inputs. The initial input rate multiplied by the ideal lead time, T_{LT} *, is changed into the ideal WIP, WIP*, which is as an input of the WIP controller.

Similar to the input variable, the output of the production system is indicated as an actual production rate, PR, and actual WIP, WIP. The difference between demand (order) and PR is integrated over a time interval, the result is the backlog (mainly refers to the backlog of order here), as another input of the backlog controller with a delay time, T_D. Thus, any demand disturbance such as rush order will be reflected on the backlog controller.

Taking production assembly line as an example, the production process is simply modeled as a pipeline where the outflow (PR) is lagged by the production lead time, T_{LT}, [24]. PR is only relevant to material stock (WIP) in the production system and the production lead time (T_{LT}). Meanwhile, assuming mean lead time is equal to the ideal lead time, T_{LT} *, thus it becomes a first-order material delay, in which material is fed and from which it flows once the material has passed through the pipe, and the sequence of material flowing into the system has no effect.

4.4.2 Design of Controllers Based on Hill Function

In a balanced production system, the input rate is set to be equal to the order rate perspective from Little's Law [25]. In other words, the production system can function in a stable state if the mean input and output match each other over a certain period of time, otherwise WIP will increase constantly due to the input rate greater than production rate, *PR*, or it will decrease due to the input rate less than *PR*. If a demand disturbance such as rush order occurs, the balance of production system is lost, and the backlog comes up on account of relatively lower capacity.

As mentioned above, a backlog controller based on Hill function is designed using control-theoretic methods. A simplified closed-loop backlog controller model is presented in Fig. 4.3. The capacity of the production system is manipulated to control backlog, with the goal of reducing any backlog resulting from the demand disturbance lagging behind the work planned. For the purpose of gaining insight into the fundamental dynamics of the production system and the effects of various choices of controllers, the following backlog model can be used:

$$B(t) = \int C_p(t)dt - \int C_a(t)dt \tag{4.4}$$

where $C_p(t)$ is the planned capacity (equal to demand here), $C_a(t)$ is the actual capacity of the production system. Taking the capacity and the backlog as two types of hormone, the level of hormone concentration can be indicated as ΔCap and ΔB, and it is supposed that the backlog hormone is controlled by the capacity hormone. If ΔB is not equal to zero, it means that unexpected disturbance occurs in the production system, and the hormone regulation will be triggered for system equilibrium through capacity regulation. The backlog controller can be described as

$$C_a(t) = \alpha F_{\text{up}}(\Delta B) + C_f(t) \tag{4.5}$$

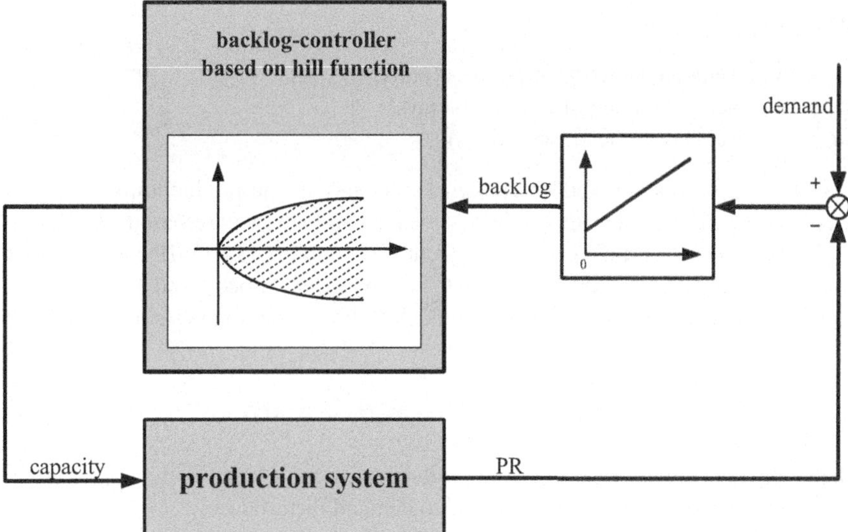

Fig. 4.3 Backlog controller based on Hill function

subject to

$$F_{\text{up}}(\Delta B) = \frac{\Delta B}{T_{bl} + \Delta B} \tag{4.6}$$

$$\Delta B = |B(t+1) - B(t)| \tag{4.7}$$

where $C_f(t)$ is the full capacity, and $F_{\text{up}}(\Delta B)$ is a backlog control function that regulates capacity by up-regulatory Hill function adaptively with the variable of backlog difference ΔB, $\Delta B > 0$. α is a positive regulation factor which is set by the backlog controller, T_{bl} is the backlog controller adjusting threshold, $T_{bl} = C_f(t) * \max(\Delta Cap)$.

In the production process, the lead time directly affects the ability for timely delivery. According to Little's Law, WIP inventory is equal to the product of the input rate and lead time. When the output rate reaches production capacity, and remains substantially constant, the amount of change in lead time is proportional to the amount of change in WIP stock, which can be expressed as Eq. (4.8). Therefore, the control of WIP inventory in production systems is equivalent to the control lead time of the production. Moreover, WIP is easier to measure by counting jobs, while lead time requires clocking jobs in and out of the system. Lead time is also restricted to its lower limits (equals to the sum of the production time and the transportation time). Therefore, it is much more practical to indirectly control lead time by controlling WIP.

$$\text{Little's Law}: \quad mwip = inr * mlt \tag{4.8}$$

where

$mwip$ represents mean work-in-process (parts)
inr represents mean input rate (parts/days)
mlt represents mean lead time (days).

WIP is accumulative, and characterizes the state of the production system and generates the information upon which decisions and actions are based. WIP gives production system inertia and provides them with memory [24]. WIP creates delays by accumulating the difference between the input rate to a process and its output rate. Based on the analysis above, the WIP model of the production system is presented as follows:

$$\text{WIP}(t) = W_{\text{in}}(t) - W_o(t) + W_d(t) \tag{4.9}$$

where $W_{\text{in}}(t)$ is the work input of the production system, $W_o(t)$ is the work output, and $W_d(t)$ represents the work input due to demand disturbances.

The main function of the WIP controller illustrated in Fig. 4.4 is to keep the production system in an operational condition. The ideal WIP, WIP*, is the reference variable. Referring to the difference between the ideal and the actual WIP, the WIP

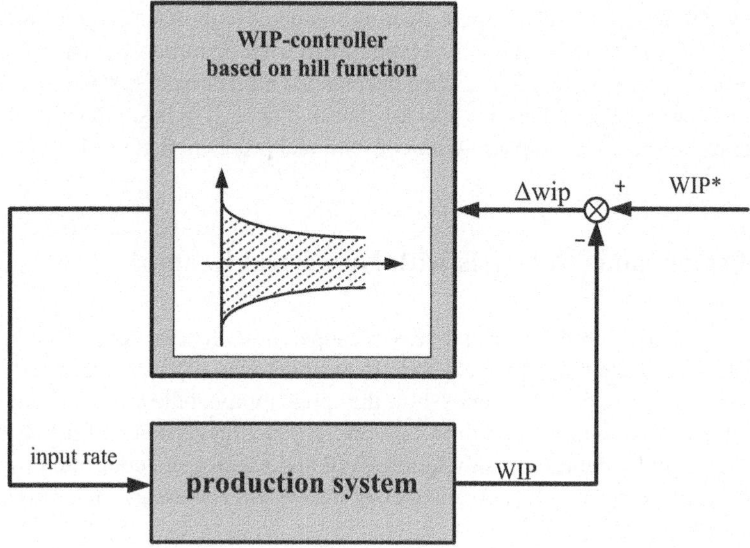

Fig. 4.4 WIP controller based on Hill function

controller adjusts the input rate of the production system. Similar to the backlog controller design, considering the input rate and the WIP as two kinds of the hormone, the deviation of hormone concentration is defined as ΔI and ΔWIP. It is supposed that the WIP hormone release is controlled by the input rate hormone. If demand disturbance occurs, the balance of the production system is broken, ΔWIP is not equal to zero. And the hormone release will be stimulated for system equilibrium through input rate hormone release. The WIP controller based on Hill function can be introduced as

$$I_a(t) = \beta F_{\text{up}}(\Delta\text{WIP}) + I_{a0} \tag{4.10}$$

subject to

$$F_{\text{up}}(\Delta\text{WIP}) = \frac{\Delta\text{WIP}}{T_{\text{WIP}} + \Delta\text{WIP}} \tag{4.11}$$

$$\Delta\text{WIP} = \left|\text{WIP}^* - \text{WIP}\right| \tag{4.12}$$

where $F_{\text{up}}(\Delta\text{WIP})$ is the regulating function of the WIP controller, ΔWIP is the difference between the current WIP value and the ideal value, $\Delta\text{WIP} > 0$, β is a small constant, I_{a0} is the initial input rate, and T_{WIP} is the adjusting threshold of the WIP controller, $T_{\text{WIP}} = \text{WIP}^*$.

The WIP controller is relatively fast-acting compared to the backlog controller and the WIP adjustment responds relatively quickly to work disturbances. The proposed backlog controller only functions, when the planned utilization of the manufacturing system is reached since otherwise backlog does not arise. The basic functionality of both controllers can be compared with conventional production control strategies.

4.5 Performance Analysis with Numerical Example

This section firstly introduces the process of cooperation between the WIP controller and backlog controller, and then explores the responses to work disturbances of the developed control model. In order to show the applicability of the proposed approach in a different situation, the lead time T_{LT}, and production capacity adjusting delay time T_D, are set by the user in simulation; WIP and backlog are controlled by their controllers, respectively. The performance of the system is measured in terms of the production rate over time.

To make the research more universal, the definition of input and output variables of the production system is made with the following assumptions:

The initial input rate, which refers to the utilization of equipment and the skill level of operators, is set to be equal to the order rate.

(1) The production system has a capacity of 40 K parts of goods every day, and the assembly line has a 95% utilization level normally (with no disturbance).
(2) When all production equipment and the operators are involved, the system runs at full capacity ($C_f(t) = 40$ K), then the production line has full utilization level.
(3) Capacity can be adjusted up to 10% of the full capacity through hard and soft activities within 2 ($T_D = 2$) days.
(4) The expected lead time is equal to the actual value ($T_{LT} = T_{LT}^* = 1.3$ days).

4.5.1 Operation of the Control Model

In the case of work disturbance occurring in a manufacturing system, it is useful to adjust the capacity by the backlog controller. If the input rate keeps growing, the WIP controller is activated to reduce the input rate of the system. Furthermore, capacity adjustments were assumed to be delayed by a time period, therefore, in order to precisely adjust capacity in a running manufacturing system, a specific reaction time (delay time, $T_D = 2$) is required.

If a demand disturbance occurs, whether or not to activate the WIP and backlog controllers, it is necessary to calculate the minimum interval for reducing the backlog firstly. Hence, the term BL_{min} (minimum backlog) is proposed, which makes a decision trigger to the controllers. The detailed process of calculating BL_{min} can be described as follows:

(1) When a demand disturbance (rush order) occurs, then the backlog comes up. The time which spends on reducing the backlog to zero using a higher capacity, ΔCap, ($\Delta Cap = 5\% * C_f$), within a balanced manufacturing system, is named as T_0, where $T_0 = BL_{min}/\Delta Cap$.

(2) Similar to the calculation above, the time used for the same backlog within the proposed model is divided into two parts. The first part (order release) is the time for the WIP adjusting process, T_1, which is relatively shorter. It refers to the required time that the WIP reaches to the ideal value, WIP*, during this period the backlog does not arise until the planned utilization is reached. The second part (order fulfillment), T_2, is the time for reducing the backlog with the hormone regulating capacity, ΔCap_{con} (according to assumptions, $\Delta Cap_{con} = 10\% * C_f$), which can be roughly calculated, here $T_2 = BL_{min}/\Delta Cap_{con}$. In fact, the actual value of T_2 is much greater than the calculated value because of the Hill functions' nonlinearity.

(3) By means of the equation of $T_0 \geq T_1 + T_2$ solved by (1) and (2), the result of BL_{min} can be calculated.

Based on the above analysis, if the actual value of the backlog is greater than the calculated BL_{min}, the WIP controller is triggered immediately, and then the backlog controller is activated when the actual value of WIP reaches the ideal value, WIP*. Figure 4.5 shows the detailed procedure of both the backlog and WIP controllers integrated into the production system. The operation of the control model is implemented as follows:

(1) The utilization of the system is improved to increase the input rate if necessary till full capacity.

(2) The WIP controller is activated to detect WIP hormone and secret the input rate hormone (order release), so that there is enough work in the system. The output (production rate) remains unchanged until the actual value, WIP, reaches the ideal (planned) value, WIP*.

(3) The backlog controller is put into operation when the actual value of WIP reaches the ideal value, WIP*, and then releases the backlog hormone to increase the capacity (secret capacity hormone) for sufficient capacity available to reduce the backlog faster.

(4) When the backlog is reduced to zero, production system achieves a balance again under coordination between the backlog controller and the WIP controller over a period of time.

4.5.2 Analysis of the Control Model Under Normal State

In order to analyze the production system with the integrated backlog controller and WIP controller, a case scenario was designed for stressing the system under the demand disturbances environment. A rush order with a working content of 90 K

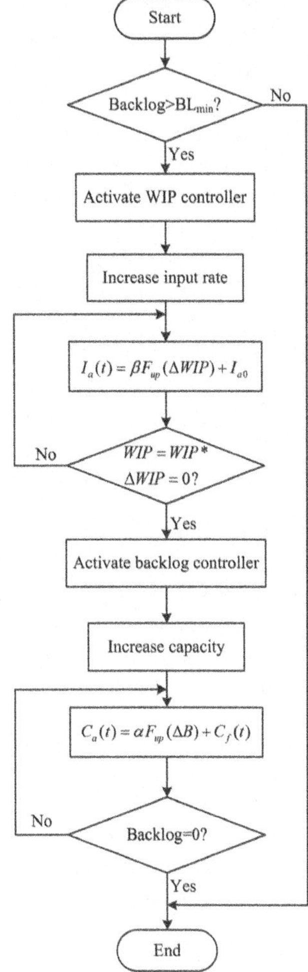

Fig. 4.5 Flow diagram of the operation procedure of the production system

pieces of goods occurs in the system on the 10th day as shown in Fig. 4.6. As an urgent order, it should be dealt with immediately after its arrival, the backlog comes up to 50 K pieces of goods because the normal planned work cannot be implemented. In the case of the rush order, the actual value of the backlog is 20 K, which exceeds the calculated value of BL_{min}. Due to this, the WIP controller increases the input rate by 38 K parts to 55 K parts in the following 6 days. Thus, there is sufficient work prepared to process for the next period within the enhanced capacity and the backlog is not reduced during this period, it demonstrates that work cannot be released and performed until the backlog controller improves enough capacity available to carry the work out. When the backlog controller is activated, the performance is enhanced up to 10%, both the backlog and WIP are smoothly reduced to the initial level relatively fast (22 days) because of its low inertia. The conventional production

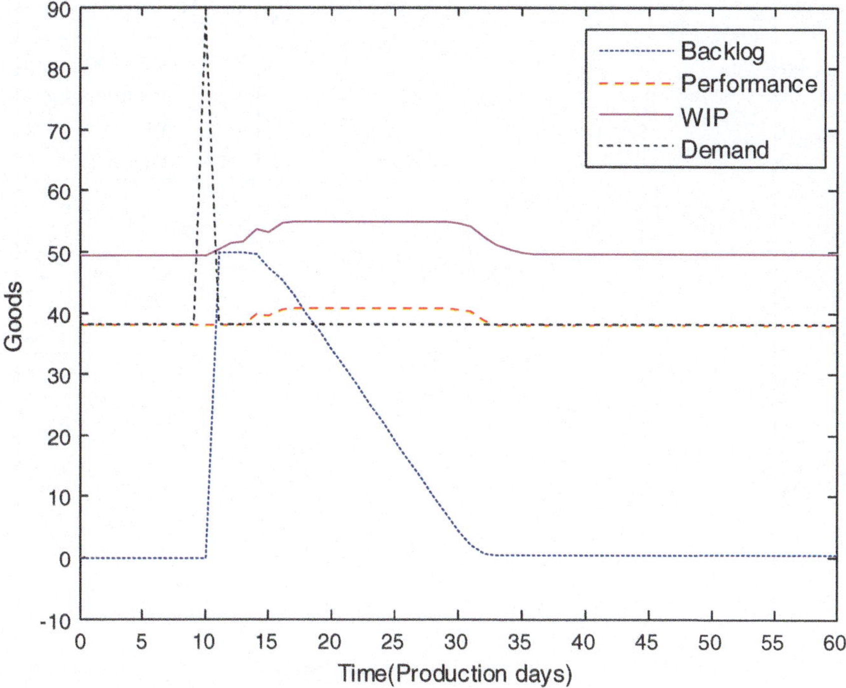

Fig. 4.6 Performance of the proposed control model under normal state

system (without WIP controller) completely reacts differently as shown in Fig. 4.7. The WIP only increases a little due to physical limit (here means that the input rate cannot be adjusted on a large scale) and the backlog reduces very slowly because of the limited capacity. In this case, the influence of WIP on backlog diminishment tends to have a main effect.

If the balanced manufacturing system was driven with a higher utilization, the conventional production system would take much more time, while the proposed control model can rapidly respond to the sudden demand change (rush order), completely eliminating the backlog without production overshoot, which has a better performance than that of the conventional manufacturing system. The greater the disturbance is, the more obvious the advantage of the control model is. The detailed analysis will be explored in the following subsection.

4.5.3 Analysis of the Control Model Under Extreme State

In the extreme case of a very high utilization of the manufacturing system with a higher level of WIP (WIP $\approx$ WIP*), $F(\Delta\text{WIP}) \approx 0$, the WIP controller tends to have a negligible effect. In this case, the backlog controller is mainly responsible for

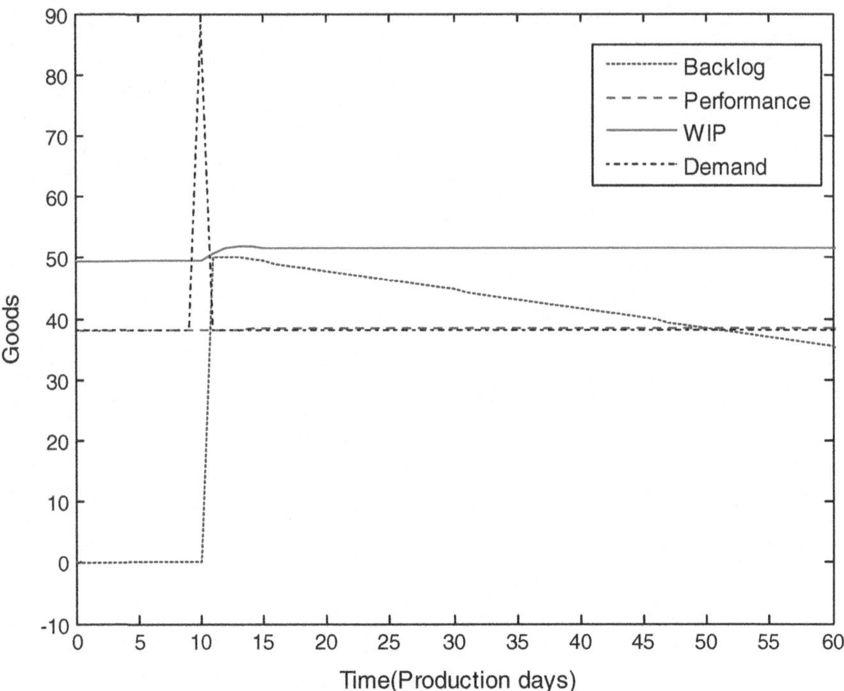

Fig. 4.7 Performance of the conventional production system under normal state

regulating tasks due to the impact of rush orders. To further illustrate the characteristics of the proposed control model, the performance of the backlog controller based on the Hill function compared with a conventional proportional backlog controller [5] is analyzed.

To examine the effect of two types of backlog controllers designed on the response of the manufacturing system, the response of both systems with the same demand changes is plotted. The same system parameters are used in the simulation. The results are shown in Figs. 4.8 and 4.9. The analysis of Fig. 4.8 shows that if a manufacturing system with a proportional backlog controller is adopted, the production process in the case of various demand disturbances (pulse responses) will appear to be frequently oscillatory under higher utilization. This problem disappeared in Fig. 4.9 due to the existence of the proposed control model inspired by hormone regulation.

How the production oscillation decreases with the hormone regulation mechanism is shown in Fig. 4.10. The result indicates that the backlog of the proposed model quickly comes up due to a sudden change in demand. That is, when the concentration of the backlog hormone increases, the capacity hormone (here the capacity hormone is equivalent to the input rate hormone) is continuously secreted driven by excitatory input of the rising backlog. The release of the capacity hormone is up-regulated by the backlog hormone. The capacity hormone itself, exerts negative nonlinear feedback on the secretion of the backlog hormone. With the decrease of the backlog hormone,

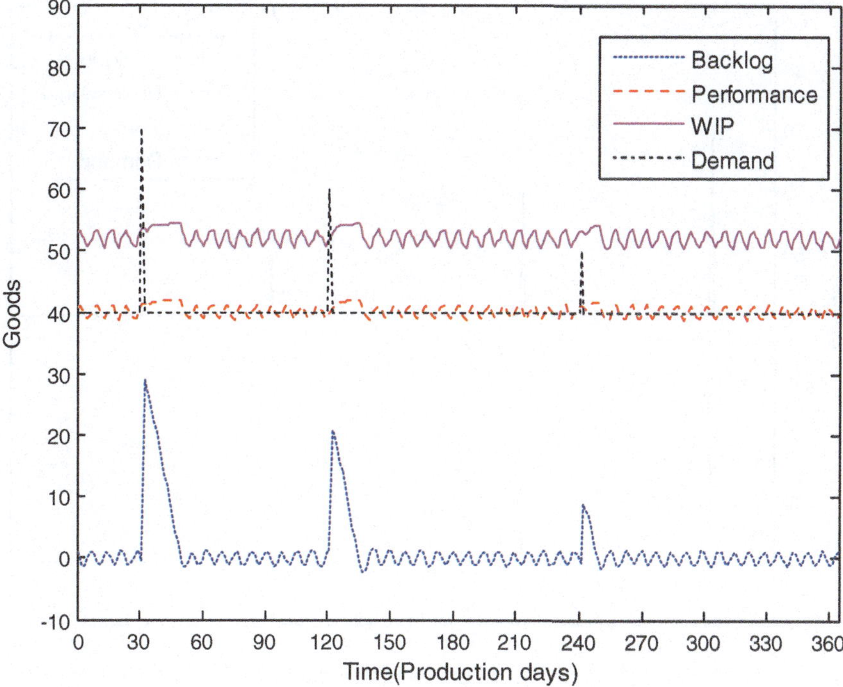

Fig. 4.8 Performance of conventional production system under extreme state

the capacity hormone rises and reaches its physical limit (action threshold), which, in turn, promotes an earlier suppression of the backlog hormone secretion. Hence, the backlog can be eliminated smoothly without production oscillation.

The analysis above reveals that the response of the production system integrated with the backlog controller based on Hill function is much better than that of the system with the conventional proportional backlog controller. Simulation results demonstrate that the proposed control model can steadily eliminate the backlog on account of the sudden demand change and make smooth regulation.

4.6   Conclusions and Future Work

Modern manufacturing systems are characterized by dynamism and customer driven based on market changes. Bio-cybernetics proves that the neuroendocrine system in the human body is quite capable of adapting to stimulus. As the control center of hormones, neuroendocrine system is featured with self-adaptation and stability, which exhibits hormone regulation behavior.

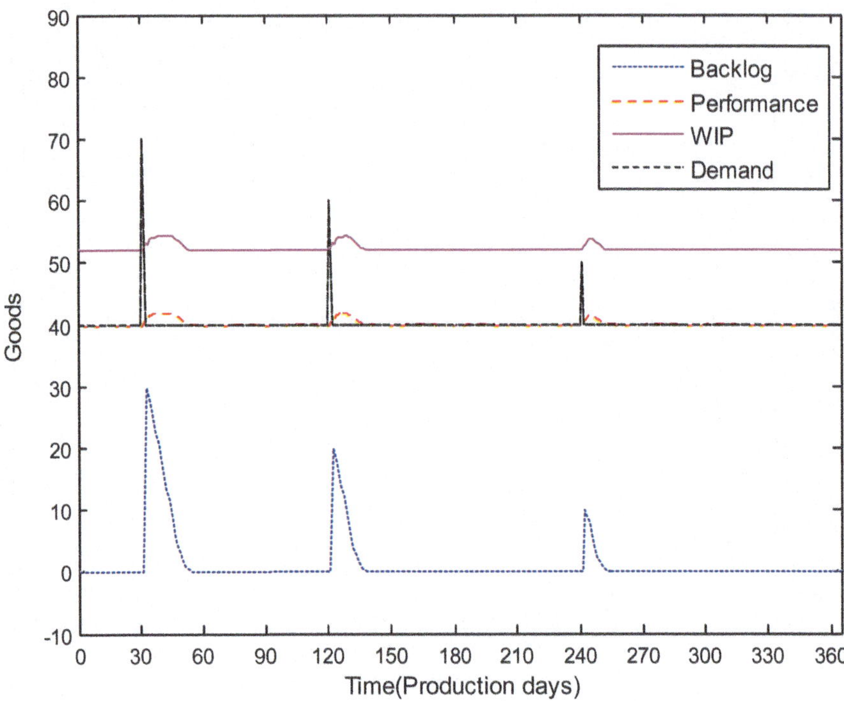

Fig. 4.9 Performance of the proposed control model under extreme state

Negative feedback regulation of hormone release is a common feature for hormones of the hypothalamic–pituitary axis. The regulation based on the Hill functions can work to maintain hormone levels within an appropriate range. In this paper, a control model of the production system was proposed and analyzed using control theory after looking into the rule of neuroendocrine hormonal regulation. The backlog and WIP control models were designed which imitate the principle of hormone regulation to agilely deal with the demand disturbances in the production system. Control laws based on the Hill functions for adjusting capacity in response to backlog and input rate as a function of desired WIP also have been studied. The simulation method was conducted to explore the performance of the proposed control model.

To examine the response of the production system under rush orders, a backlog controller and a WIP controller with an exponential delay component were designed. Simulation results showed that the developed model with backlog and WIP controllers based on the Hill functions which can adjust control parameters adaptively in real time, is more responsive than the conventional production system. The backlog was reduced steadily and much faster while the WIP was maintained within the desired level. To further investigate the developed backlog and WIP controllers, simulation analysis of extreme state related to much higher utilization of

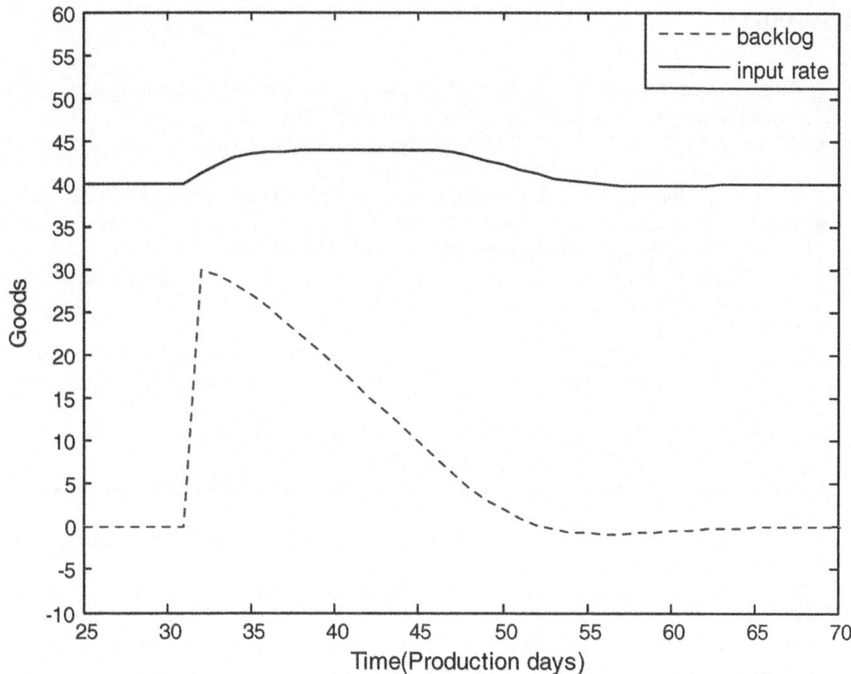

Fig. 4.10 Mutual interactions between the backlog hormone and the input rate hormone

the manufacturing system between the proposed backlog controller and the conventional proportional backlog controller was implemented. Results of comparing the two types of backlog controllers showed higher robustness to rush orders when adopting the control approach with hormone regulation. Also, the developed controllers decreased the overshooting of production due to the Hill functions' nonlinear characteristics during the backlog reducing process. This confirms the known fact that hormone regulation is a better optimal adjusting method. It was also shown that the production oscillation problem can be better managed with the proposed model which makes a good balance between demand responsiveness and stable production performance.

Future work is required to investigate the effect of different setting values of the control parameter, especially T_{bl} and T_{WIP}, in the developed control model. An effort should be made to detail the input rate and capacity adjusting procedures when adopting hormone regulation. Further research on the application of hormone regulation should be studied. Also, the response of the production system to other kinds of demand disturbance or internal turbulence needs to be tested. Finally, more theoretical analyses are desired to be properly mapped to real practice.

References

1. Ueda, K., Vaario, J., & Ohkura, K. (1997). Modeling of biological manufacturing systems for dynamic reconfiguration. *Annals of the CIRP, 46*(1), 343–346.
2. Ueda, K., Kito, T., & Fujii, N. (2006). Modeling biological manufacturing systems with bounded-rational agents. *Annals of the CIRP, 55*(1), 469–472.
3. Chan, F. T. S., Tibrewal, R. K., & Prakash, A. (2015). A biased random key genetic algorithm approach for inventory-based multi-item lot-sizing problem. *Proceedings of the Institution of Mechanical Engineers, Part B: Journal of Engineering Manufacture, 229*(1), 157–171.
4. Tang, D. B., Gu, W. B., Wang, L., & Zheng, K. (2011). A neuroendocrine-inspired approach for adaptive manufacturing system control. *International Journal of Production Research, 49*(5), 1256–1258.
5. Deif, A. M., & EIMaraghy, W. H. (2006). A control approach to explore the dynamics of capacity scalability in reconfigurable manufacturing systems. *Journal of Manufacturing Systems, 25*(1), 12–18.
6. Kim, J. H., & Duffie, N. A. (2004). Backlog control design for a closed loop PPC system. *CIRP Annals-Manufacturing Technology, 53*(1), 357–359.
7. Duffie, N., & Falu, I. (2002). Control-theoretic analysis of a closed loop PPC system. *Annals of the CIRP, 52*(1), 379–382.
8. Nyhuis, F., & Pereira Filho, N. A. (2002). Methods and tools for dynamic capacity planning and control. *Gestao Producao 9*(3), 245–260.
9. Farhy, L. S. (2004). Modeling of oscillations of endocrine networks with feedback. *Methods in Enzymology, 384,* 54–81.
10. Evans, W. S., Farhy, L. S., & Johnson, M. L. (2009). Biomathematical modeling of pulsatile hormone secretion: a historical perspective. *Methods in Enzymology, 454*(1), 345–366.
11. Zhang, H. T., Tang, D. B., et al. (2016). Production control strategy inspired by neuroendocrine regulation. *Proceedings of the Institution of Mechanical Engineers, Part B: Journal of Engineering Manufacture, 232*(1), 67–77.
12. Towill, D. (1982). Dynamic analysis of an inventory control and order-based production control system. *International Journal of Production Research, 20,* 671–687.
13. Qiu, M. M., Fredendall, L. D., & Zhu, Z. (2001). Application of hierarchical production planning in a multimachine environment. *International Journal of Production Research, 39*(13), 2803–2816.
14. Asl, R., & Ulsory, A. (2002). Capacity management via feedback control in reconfigurable manufacturing systems. In *Proceedings of Japan-USA Symposium on Flexible Mfg. Automation 2002,* Hiroshima, Japan.
15. Deif, A. M., & EIMaraghy, W. H. (2005). Capacity scalability in reconfigurable manufacturing systems. In *Proceedings of CIRP 3rd Conference for Reconfigurable Manufacturing Systems,* 2005.
16. Guo, C. F., & Jing, R. Z. (2011). Modeling and controlling work-in-progress in discrete manufacturing systems. *Kybernetes, 40*(5), 842–846.
17. Wiendahl, H. P. (1995). *Load-oriented manufacturing control.* Hannover, NH: Springer.
18. Wiendahl, H. P., & Breithaupt, J. (2000). Automatic production control applying control theory. *International Journal of Production Economics, 63*(1), 33–46.
19. Wiendahl, H. P., & Breithaupt, J. (2001). Backlog-oriented automatic production control. *Annals of the CIRP, 43*(2), 533–540.
20. Nyhuis, P. (1994). Logistic operating curves—A comprehensive method for rating logistic potentials. *EURO XIII/OR36* 1994, University of Strathclyde, Glasgow.
21. Savino, W., & Dardenne, M. (2000). Neuroendocrine control of thymus physiology. *Endocrine Reviews, 21*(4), 412–443.
22. Liu, B., Ren, L., & Ding, Y. S. (2005). A novel intelligent controller based on modulation of neuroendocrine system. In *Proceedings of the 2nd International Symposium on Neural Networks: Advances in Neural Networks* (pp. 119–124) Springer.

23. Guo, C. B., Hao, K. G., & Ding, Y. S. (2012). Neuroendocrine-based cooperative intelligent control system for multi-objective integrated control of a parallel manipulator. *Mathematical Problem in Engineering, 20*(6), 1–17.
24. Sterm, J. D. (2000). *Business dynamics: Systems thinking and modeling for a complex world.* New York: McGraw-Hill.
25. Little, J. D. C. (1961). A proof for the queuing formula $L = \lambda W$. *Operations Research, 9,* 383–387.

Chapter 5
Neuroendocrine-Immune Regulation Based Approach for Disturbance Handling

5.1 Introduction and Synopsis

Disturbances in manufacturing systems are mainly divided into two categories: external disturbances and internal disturbances [1–7]. External disturbances include changes in tasks and orders due to dynamic market environment (emergency orders, task additions, mission cancelations, etc.), process changes caused by customized requirements, supply of raw materials, etc. Internal disturbances include equipment failure and maintenance, personnel absence, tool damage, production delay, operational errors, etc. The occurrence of these uncertain disturbances often results in a failure of the original plan or makes the target impossible to accomplish [8]. Disturbances will lead to the decline of manufacturing system productivity and miss business opportunities, thus losing market competitiveness [9].

Traditional manufacturing systems rely on centralized control architecture with the advantage of the most optimized production plan, but due to the limitations of centralized architecture, the ability of responsiveness to disturbances is insufficient. When a disturbance occurs, the operation of whole manufacturing systems is interrupted, rescheduling of the whole production plan is needed to restart production. Therefore, quick response and disturbance handling is a research hotspot of the Intelligent Manufacturing System. Holon and Agent approaches have been one direction for disturbance handling in the past decade. With autonomous and negotiated characteristics, Holon and Agent approaches have great advantages in dealing with disturbances compared to traditional manufacturing systems with centralized control architecture. Merdan et al. [10] investigated the dynamic scheduling of the manufacturing system under the disturbance environment by using the multi-agent method and obtained better performance indicators. In order to avoid influences caused by disturbances, multi-agent-based "proactive-reactive" scheduling method was proposed, which shows better ability in job shop scheduling problem [11]. The proactive scheduling generated a robust predictive scheduling plan to deal with the disturbance; and the reactive scheduling dynamically corrected the

© Springer Nature Singapore Pte Ltd. 2020
D. Tang et al., *Adaptive Control of Bio-Inspired Manufacturing Systems*,
Research on Intelligent Manufacturing,
https://doi.org/10.1007/978-981-15-3445-4_5

predictive scheduling plan according to the disturbance. Leitão [12] introduced a disturbance handling architecture for holonic manufacturing system (HMS), which uses distributed features of HMS to deal with disturbances and adjust scheduling by rescheduling approach. Jana et al. [13] adopted an agent-based adaptive scheduling approach to handle machine breakdown in HMS architecture. Under normal circumstances, task holons and resource holons obtain a scheduling plan by coordinating with the integrated scheduling holon. In the case of a machine breakdown, tasks from faulty machine are assigned by task redistribution mechanism inspired by human cognitive behavior.

For disturbance handling of manufacturing systems, many researchers have investigated dynamic scheduling implementation of operation tasks. However, very few studies involve the scheduling of transportation tasks. Especially in a dynamic and real-time environment, few researchers consider the simultaneous scheduling of operation task and transportation task. Umar et al. [14] develop a multi-objective genetic algorithm to deal with "dynamic scheduling of operation task" and "AGV dispatching" in the flexible manufacturing system. The algorithm generates a comprehensive scheduling plan and detailed paths and optimizes the completion time of the whole system. This algorithm is an offline approach that is difficult to adapt disturbances in a dynamic and real-time environment. Erol et al. [15] proposed a multi-agent system (MAS) approach to carry out the scheduling of operation tasks and AGVs in real-time environments. The approach uses the bidding and negotiation between Agents to generate feasible scheduling solutions in real time, but the influence of disturbance is still not taken into consideration.

In this chapter, inspired by the neuroendocrine-immune mechanism, a disturbance handling approach was proposed in the architecture of BIMS. This approach can quickly detect various disturbances in the manufacturing system, then combine dynamic shop floor scheduling approach and AGV on-line scheduling approach to quickly handle disturbances and optimize task and resource allocation.

5.2 Disturbance Handling of BIMS

5.2.1 Disturbance Handling Mechanism of BIMS

Based on neuroendocrine-immune mechanism, a disturbance handling mechanism of BIMS is shown in Fig. 5.1. The mechanism consists of three functional modules listed as follows:

Immune monitoring function: monitor the state of BIMS in real time, and rapidly identify disturbances in the case of abnormal factors, then generate strategies (antibodies) for BIMS to achieve rapid response to disturbances. This function involves all BIMCs of BIMS with its own immune monitoring module, which can perform immune monitoring.

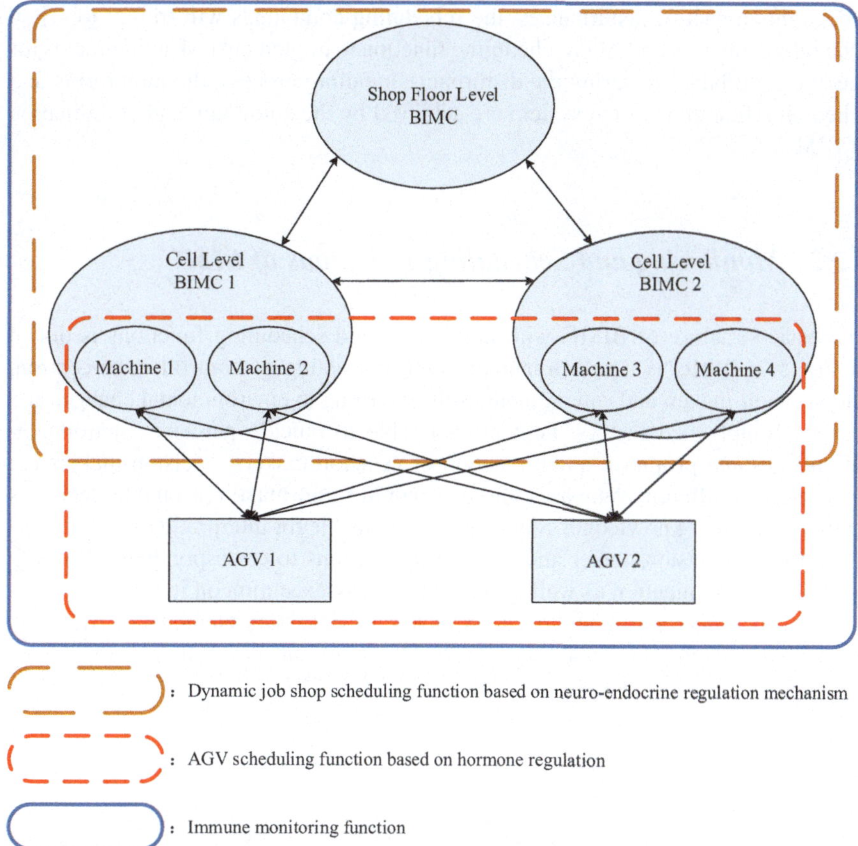

Fig. 5.1 Disturbance handling mechanism of BIMS

Dynamic job shop scheduling function based on neuroendocrine regulation mechanism: deal with dynamic allocation of operation tasks between machines under disturbance environment [16].

AGV scheduling function based on hormone regulation: realizes the distribution of transportation tasks between AGV and machine tools in on-line environment [17].

The system consists of a shop floor level BIMC, cell level BIMCs, machine-level BIMCs, and AGV-level BIMCs. Disturbance handling of BIMS is accomplished by neuroendocrine-immune coordination between BIMCs at different levels.

When unexpected disturbances occur in the system (such as machine failures, emergency orders, AGV failures, delays, etc.), BIMS can maintain the stability of the manufacturing system based on neuroendocrine-immune mechanism. BIMCs from different levels adopt the immune monitoring function to monitor the manufacturing system in real time, then identify and diagnose the disturbance according to the immune response mechanism to generate new scheduling commands. To reduce the impact of disturbances, the scheduling commands will trigger job shop scheduling function and AGV scheduling function to implement task assignments for specific disturbances. During the disturbance handling process, the monitoring and scheduling functions of the system are achieved by the autonomy and coordination of BIMCs.

5.2.2 Monitoring and Scheduling Functions of BIMC

The basic structure of BIMC with monitoring and scheduling functions is drawn in Fig. 5.2. BIMC is a self-organizing unit, consisting of controller, perceptron, and decision-maker, and can regulate itself according to environmental changes and various complicated factors. Perceptron is able to quickly perceive environmental changes and prompt a quick reaction of decision-makers; decision-makers are connected to different databases and can monitor and make reasonable decisions according to their knowledge; controller is responsible for interpreting the decisions made by the decision-maker and sending instructions to corresponding BIMCs to establish communication as well as performing task execution on itself.

Controller, perceptron, and decision-maker are necessary elements of BIMC, in which the decision-maker plays a core role. Decision-makers can analyze the impact

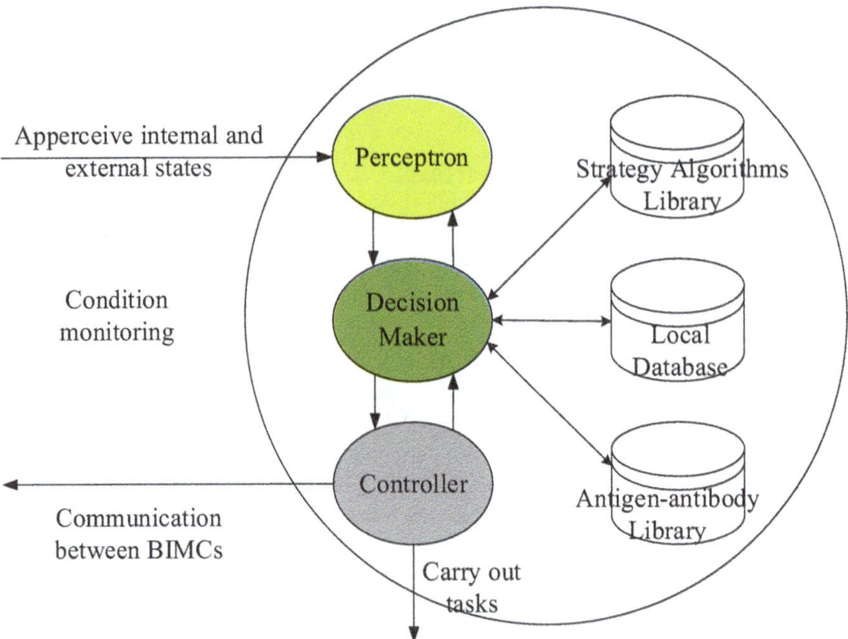

Fig. 5.2 Architecture of BIMC with monitoring and scheduling functions

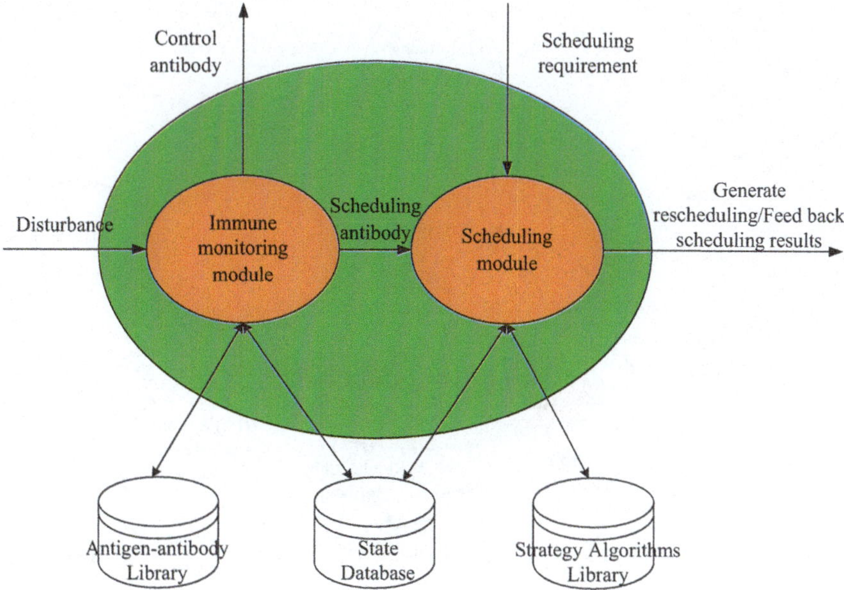

Fig. 5.3 Decision-maker based on neuroendocrine-immune coordination

of disturbances, formulate corresponding strategies, and initiate dynamic scheduling. According to the above functions, the decision-maker architecture, including the immune monitoring module and scheduling module, is shown in Fig. 5.3. The immune monitoring module conducts the detection and diagnosis of manufacturing system status. When the disturbance occurs, it can be quickly detected and the corresponding immune response will be reacted to generate an antibody which is consisted of the control antibody and scheduling antibody. Control antibody acts on the controller to execute the control command, and scheduling antibody acts on the scheduling module to trigger dynamic scheduling. Scheduling module integrates different kinds of scheduling algorithms (transport task scheduling based on hormone regulation mechanism, operation scheduling based on neuroendocrine regulation mechanism, etc.), which can cope with various scheduling requirements.

5.2.3 Disturbance Handling Processes of BIMC

Due to local information limitations, BIMC can use the embedded algorithms and knowledge to deal with disturbances, but they do not have enough capability and knowledge to perform this work alone. Therefore, all disturbances are handled in a distributed manner through the cooperation of different levels of BIMCs. Figure 5.4 shows the disturbance handling processes of BIMC.

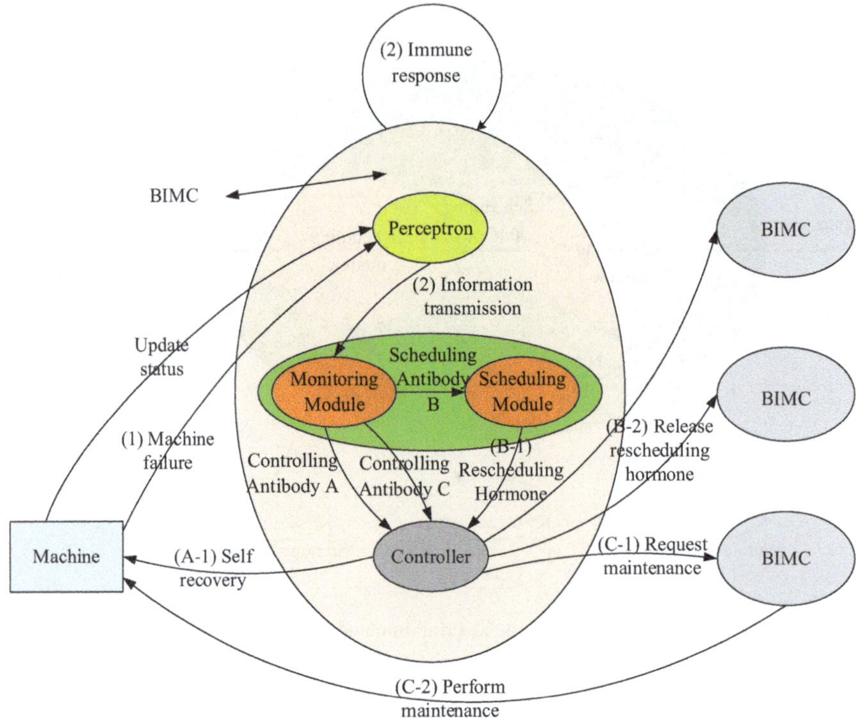

Fig. 5.4 Disturbance handling processes of BIMC

BIMCs at each level monitor the operating status of the manufacturing system in real time and compare with normal status to identify deviation. Once the deviation (e.g., machine failure) is detected by a BIMC, the BIMC will perform an immune response operation, generate antibodies (A, B, C), and indicate possible operations to be performed for manufacturing system recovery. If an operation cannot be solved internally by the BIMC itself such as scheduling antibody B and controlling antibody C, it needs to be coordinated with other BIMCs. In order to handle disturbances, BIMS implement disturbance detection, diagnosis, and handling strategies according to autonomy within BIMC and coordination between different levels of BIMCs. In the following, disturbances detection, diagnosis and handling strategy of BIMS will be described in detail.

5.3 Disturbance Detection and Diagnosis of BIMS

Disturbances in the manufacturing system can also be thought of as activities that deviate from the established plan and bring negative impacts on the production process. Common disturbances in manufacturing systems include machine failures,

AGV failures, emergency orders, delays, and the absence of personnel, which often bring large impacts on the manufacturing system planning and control. To handling disturbances, BIMS should quickly detect and diagnose disturbances, then choose the most appropriate approach, to ensure continuous operation of the manufacturing system and improve the robustness and productivity.

5.3.1 Disturbance Detection

In BIMS, the detection of disturbances is achieved by activities from various levels of BIMCs. Each BIMC continuously monitors the entire process of production planning and plays different roles during the monitoring process.

BIMC in the shop floor level performs inspections of the system and monitors information from other BIMCs, including scheduling plan implementation, real-time production status, health status, and external order arrivals.

BIMC in cell level performs internal detection of cell and aggregates information of BIMCs within cell as well as information from other BIMCs, including scheduling plan implementation within cell, real-time production status and health status within cell, and requirement information from other BIMC.

Monitoring function of BIMC in machine level consists of task detection, quality detection, and resource status detection. Task detection function inspects orders completed by machine tools, observes the order of executing operations and time nodes of finished tasks; quality detection function inspects the quality of production processes; resource detection function inspects the running and health status of machine tools in real time.

Monitoring function of BIMC in AGV level consists of task detection and AGV status detection. The task detection function inspects the transport tasks completed by the AGV, observes the order of transportation and time node of AGV operation; AGV status detection function inspects the operation status, health status, and location of AGV in real time.

Disturbance detection model utilizes the distributed characteristic of BIMS to implement monitoring from bottom to top. The detection process involves the acquisition and classification of information, and information evaluation is realized by the diagnostic function of disturbances.

5.3.2 Diagnosis of Disturbances

Information diagnosis in BIMS is accomplished by the immune response of the immune monitoring module in BIMC shown in Fig. 5.5. The response of immune monitoring module in BIMC includes state vector module, conventional reaction module, antigen generating module, immune antigen evaluation and recognition module, antibody-matching module, and learning module.

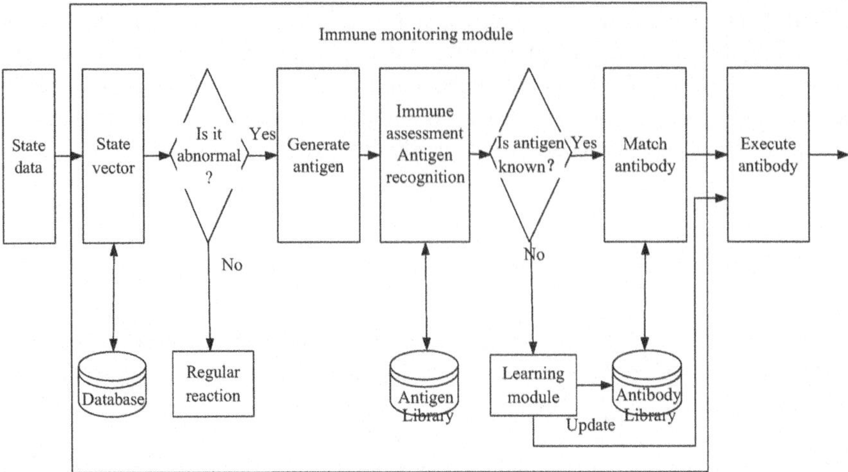

Fig. 5.5 Immune response of monitoring module in BIMC

BIMCs in different levels detect and collect the state information of the manufacturing system, and the monitoring module of BIMC manages and filters information to ensure accuracy and availability of information. In state information, only a few of them contain disturbance information of the manufacturing system. State vector module handles this state information, and transmits it into a state vector, then compares it with the vector in normal state. If the state vector is judged to be normal, the monitoring module performs a regular reaction; otherwise, it indicates that there is a disturbance in the system and the state vector is marked as a disturbed vector. The antigen generation module extracts the disturbance information from the vector and generates the corresponding antigen. Different classes of disturbances constitute antigen library of the manufacturing system. After generating of antigen, immune evaluation and recognition module evaluates the antigen. When the immune index is below the threshold, it indicates that the antigen brings a big impact on the manufacturing system. At this time, the monitoring module performs antigen recognition based on the existing antigen library. For known antigens, the antibody-matching module produces antibodies that drive the manufacturing system to quickly and accurately eliminate the effects of antigens. For new antigens, BIMS will use a learning module to generate new antibodies in special cases by artificial assistance and eliminate the effects of antigens, then update new antibodies into the antibody library.

When an antibody command is issued to relevant BIMCs, different BIMCs adopt a scheduling module to realize distributed and dynamic resource allocation and eliminate the antigen. In this way, disturbances can be controlled or weakened and the manufacturing system can be brought into a steady state again.

5.4 Disturbance Handling Strategies of BIMS

In production, the manufacturing system will be disturbed by unpredictable events. This section will take machine failure as an example to analyze the BIMS disturbance handling strategy. For convenience and description, the rest of this section will use shop floor, cell, machine, and AGV instead of BIMCs in relevant levels to describe disturbance handling strategies. Relevant symbols are defined as follows:

TO operation task;
TT transportation task;
H hormone secretion concentration of machine stimulated by transportation task;
S hormone secretion rate of AGV stimulated by transportation task;
ρ weighted value hormones for operation task;
p scheduling plan;
I cell set;
M machine set;
K AGV set;
J job set;
L operation set;
Δd deviation of due date.

When a machine failure occurs, disturbance can be identified by the immune monitoring function of the related machine, and corresponding strategies (antibodies) are given. At the same time, deviation of tasks and resources caused by machine failure is regarded as an oscillation of hormone and stimulates neuroendocrine-immune regulation of the manufacturing system. As shown in Fig. 5.6, BIMCs at all levels deal with the disturbances in case of machine failure according to the following steps:

Step (1)   When machine failure occurs, the machine m' generates corresponding strategies {(A), (B), (C), (D), (E)} according to immune evaluation and response. Strategy (A) is self-diagnosis and self-repair of machine; Strategy (B) allocates unfinished operation tasks in the task list of the machine m'; Strategy (C) and Strategy (D) allocate unfinished transportation task; Strategy (E) requests maintenance.

Strategy (A) will be executed as follows:

Step (A-1)   Machine m' performs self-diagnosis and self-repair operation.

Strategy (B) will be executed as follows:

Step (B-1)   Machine m' predicts its own state and estimates repair time, and extracts operating information $\{TO_{jl}\}$ that cannot be completed during the repair time, then feedback to its own cell i'.

Step (B-2)   Cell i' releases task (TO_{jl}) into shop floor environment as CRH.

Step (B-3)   When each cell senses a task (TO_{jl}) from Cell i', it will release TO_{jl} into its relevant cell environment as ATCH.

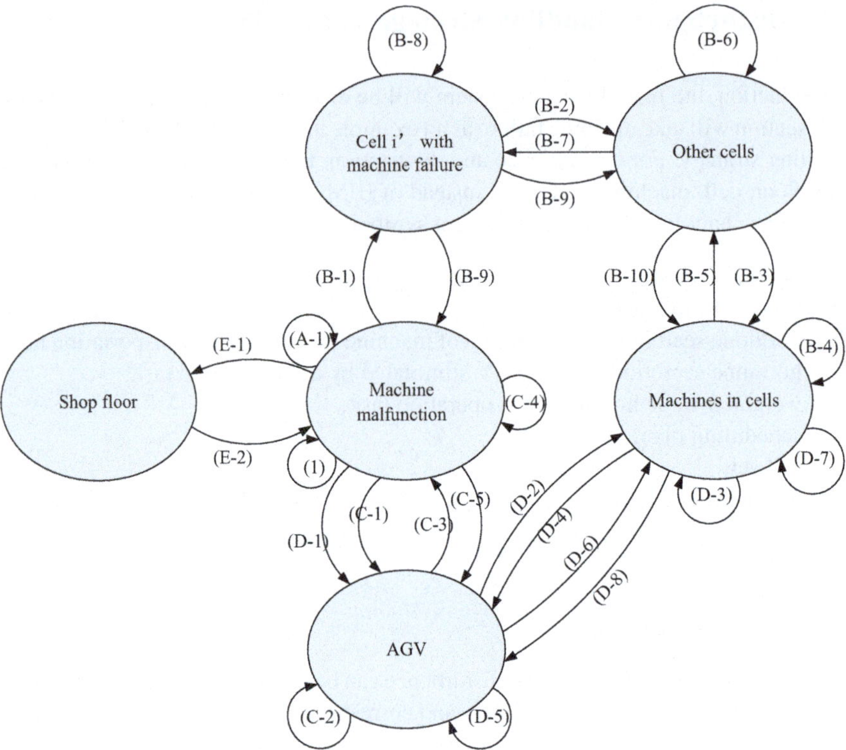

Fig. 5.6 Disturbance handling strategy for machine failure

Step (B-4)   When a machine senses TO_{jl}, it will check machine skills. If the machine has the skill to operate TO_{jl}, it will reschedule the original plan as follows. Firstly, the machine will try to insert TO_{jl} to the original plan, then evaluate the increase of hormone quantum ($\rho_{im}^{TO_{jl}}$); then the decision-maker of the machine selects the plan (p_{im}) with the minimum increase of hormone quantum from alternative production plans as the optimal plan. If the machine does not have the skill to operate TO_{jl}, it will do nothing.

Step (B-5)   Machine feeds back the new plan with the increase of hormone quantum ($p_{im}, \rho_{im}^{TO_{jl}}$) to its cell as cortical.

Step (B-6)   When the cell senses the feedback information $\{(p_{im}, \rho_{im}^{TO_{jl}})\}$ from relevant machines, it will reschedule the plan, and select the optimal plan with the minimum increase of hormone quantum ($p_i, \rho_i^{TO_{jl}}$) from alternative production plans.

Step (B-7)   The cell feeds back the new plan with the increase of hormone quantum ($p_i, \rho_i^{TO_{jl}}$) to cell i' as cortical.

Step (B-8)   When Cell i' senses the feedback information $(p_i, \rho_i^{TO_{jl}})$, it will select the optimal plan $(p_i^{opt}, \rho_i^{TO_{jl}})$ with the minimum increase of hormone quantum from alternative production plans.

Step (B-9)   Cell i' allocates TO_{jl} to relevant cell, and execute Step (B-10); The release information of TO_{jl} is informed to failure machine, and then execute Step (C-1).

Step (B-10)   cell i' allocates TO_{jl} to relevant machine.

Step (B-11)   Execute Step (B-2) to Step (B-10) cyclically, until the assignment of $\{T_{jl}\}$ is completed.

Strategy (C) will be executed as follows:

Step (C-1)   If workpiece j is in machine m', the AGV is now required to transport it to the assigned machine. Stimulated by TT_{jl}, machine m' secretes the hormone $(H_m^{jl}(t))$, releases hormone information $(H_{M'}^{TT_{jl}}, TT_{jl})$ into the shop floor environment; otherwise, go to Step (D-1).

Step (C-2)   Stimulated by $(H_m^{jl}(t), TT')$, AGV calculates the hormone secretion speed $(S_{AGV}^{TT_{jl}})$.

Step (C-3)   AGV feeds back $S_{AGV}^{TT_{jl}}$ to machine m'.

Step (C-4)   Machine m' receives the feedback information $\{S_{AGV}^{TT_{jl}}\}$ and selects suitable AGV.

Step (C-5)   Machine m' allocates TT_{jl} to relevant AGV.

Strategy (D) will be executed as follows:

Step (D-1)   If workpiece j is neither in machine m' nor in the task list of AGVs, machine m' will send transportation task $(TT^{m'})$ return instruction to AGV; otherwise, go to Step (E-1).

Step (D-2)   If the AGV is transporting workpiece j, it will continue to complete the transportation task and return it to the destination machine; otherwise, AGV will return $TT^{m'}$ to the destination machine directly.

Step (D-3)   When the machine receives the returned transportation task, it will secrete hormone $(H_M^{TT^{m'}})$ and search a new AGV for transportation.

Step (D-4)   Machine release hormone $(H_M^{TT^{m'}}, TT^{m'})$ into the shop floor environment and the transportation task will compete with other transportation missions for AGV.

Step (D-5)   Once $TT^{m'}$ is selected by an AGV, AGV will generate $(S_{AGV}^{TT^{m'}})$.

Step (D-6)   AGV feeds back $S_{AGV}^{TT^{m'}}$ to the relevant machine.

Step (D-7)   When the machine receives the feedback information set $\{S_{AGV_k}^{TT^{m'}}\}$, it will select the right AGV.

Step (D-8)   The machine allocates $TT^{m'}$ to relevant AGV, then executes Step (E-1).

Strategy (E) will be executed as follows:

Step (E-1)   Machine sends a maintenance request to the shop floor controller.

Step (E-2)   Machine implements maintenance.

During the disturbance handling of BIMS, immune monitoring mechanism is used for the rapid detection of disturbances, and giving relevant strategies; the dynamic scheduling approaches of BIMS are applied for the allocation of operation tasks and transportation tasks. On this basis, production deviations caused by disturbances can be eliminated, and the impact of disturbances is reduced. The proposed disturbance handling approach of BIMS will be verified in the next section.

5.5 Case Study

5.5.1 Experimental Description

This chapter proposes a disturbance handling approach based on neuroendocrine-immune regulation mechanism for BIMS. In order to verify the feasibility and effectiveness of the approach and obtain comparative experiments, experiments with different manufacturing system models in different scenarios are designed according to the benchmarking framework defined by Cavalieri [18].

The experiments make the following assumptions:

(1) Before an operation task is completed, the machine cannot operate another task. Each machine has enough buffers.
(2) One process of a workpiece must be performed after the previous process is completed.
(3) Setup time and post-processing time will not be considered; each machine has sufficient capacity to operate the required tasks.
(4) Operation time of task in machine and transportation time of AGV are determined.
(5) Each AGV can only transport one task at a time.
(6) Pre-emption is not allowed in transportation and operation, i.e., once task transportation or operation is started, it must be finished without interruption.

In the experiment, there are 4 machines, 2 AGVs, 4 set of orders. Each set of orders contains 5–6 tasks, and each task contains 2–4 processes which can be found in Table 5.1. The transportation time of AGV between machines and warehouse is shown in matrix (5.1).

The experiment uses JAVA as simulation environment, and adopts Windows 7 as the operating system: the processor is Intel Core Duo 2.0 GHz and 4 GB of memory. The experiment is performed in two steps. Firstly, verify the feasibility of the disturbances handling approach in different events such as machine failure, AGV failure, rush order, and delay. Then evaluate the performance parameters of the manufacturing system after running multiple tests in different scenarios. The specific shop floor scenarios are as follows:

Table 5.1 Order and task information

Order 1	Order 2
Task 1: M1(10); M4(18);	Task 1: M4(11); M1(10); M2(7)
Task 2: M2(10); M4(18)	Task 2: M3(12); M2(10); M4(8)
Task 3: M1(10); M3(20)	Task 3: M2(7); M3(10); M1(9); M3(8)
Task 4: M2(10); M3(15); M4(12)	Task 4: M2(7); M4(8); M1(12); M2(6)
Task 5: M1(10); M2(15); M4(12)	Task 5: M1(9); M2(7); M4(8); M2(10); M3(8)
Task 6: M1(10); M2(15); M3(12)	
Order 3	Order 4
Task 1: M1(9); M2(11); M4(7)	Task 1: M1(11); M3(19); M2(16); M4(13)
Task 2: M1(19); M2(20); M4(13)	Task 2: M2(21); M3(16); M4(14)
Task 3: M2(14); M3(20); M4(9)	Task 3: M3(8); M2(10); M1(14); M4(9)
Task 4: M2(14); M3(20); M4(9)	Task 4: M2(13); M3(20); M4(10)
Task 5: M1(11); M3(16); M4(8)	Task 5: M1(9); M3(16); M4(18)
Task 6: M1(10); M3(12); M4(10)	Task 6: M2(19); M1(21); M3(11); M4(15)

(1) No disturbances occur;
(2) One of M2, M3, and M4 has a 20% probability of failure;
(3) An AGV has a 20% probability of failure;
(4) There is a 20% probability of rush orders.

$$D = \begin{bmatrix} & AS/RS & M1 & M2 & M3 & M4 \\ AS/RS & 0 & 2 & 4 & 10 & 12 \\ M1 & 12 & 0 & 2 & 8 & 10 \\ M2 & 10 & 12 & 0 & 6 & 8 \\ M3 & 4 & 6 & 8 & 0 & 2 \\ M4 & 2 & 4 & 6 & 12 & 0 \end{bmatrix} \tag{5.1}$$

5.5.2 *Experiment Analysis*

In order to verify the feasibility of the disturbance handling approach, three experiments were performed on the introduction of the disturbance in order 1. Experiment 1 sets M2 to failure state at $t = 20$ s and recovers at $t = 40$ s; experiment 2 sets AGV2 to failure state at $t = 30$ s and recovers at $t = 60$ s; experiment 3 sets the introduction of a rush order set at $t = 40$ s (order 3). The Gantt charts obtained in the experiments are shown in Figs. 5.7, 5.8 and 5.9.

As shown in Fig. 5.7, M2 fails at $t = 30$ s. By using the immune monitoring module, M2 quickly detected the disturbance, and adopted neuroendocrine-immune regulation of BIMS to trigger dynamic scheduling and machine repair operations.

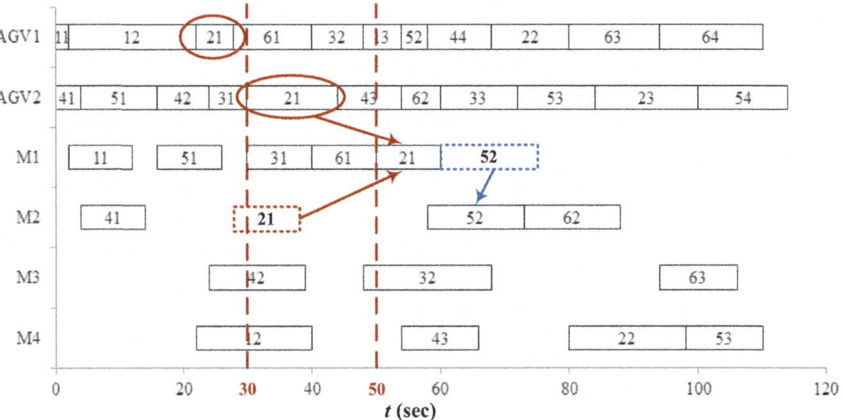

Fig. 5.7 Gantt chart when machine fault occurs

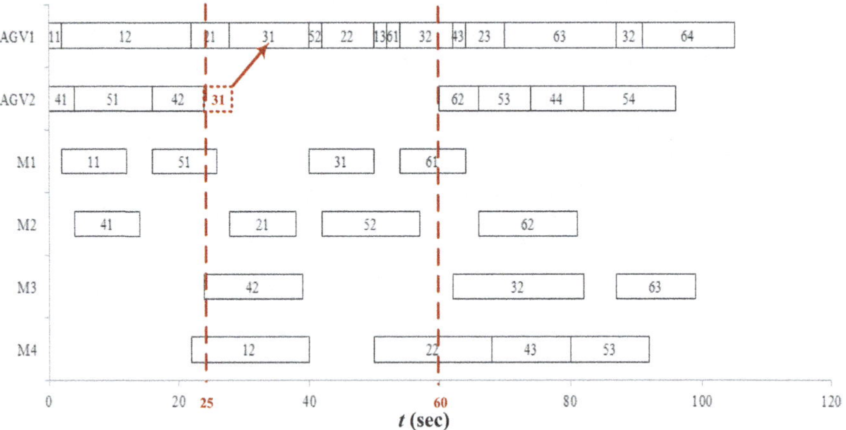

Fig. 5.8 Gantt chart when AGV failure occurs

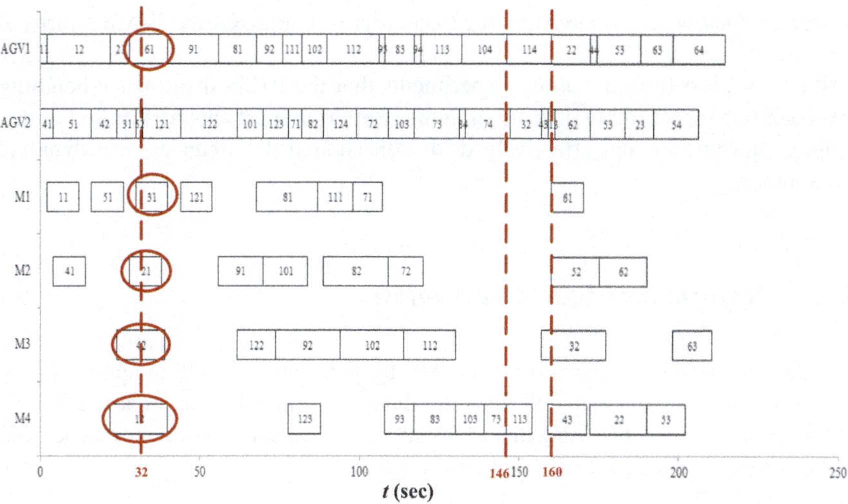

Fig. 5.9 Gantt chart when a set of rush order occurs

After rescheduling, operation task 21 and operation task 52 were assigned to M1, and the disturbance caused by machine faults at this moment was temporarily eliminated. When M2 recovered at $t =50$ s, by using immune monitoring module, M1 detected production delay (process 52), and triggered a new rescheduling to distribute operation task 52 back to M2. In this case, BIMS eliminates the disturbance and returns to the normal production state.

As shown in Fig. 5.8 AGV2 failed at $t =30$ s. By using immune monitoring function AGV2 quickly detected the disturbance, and adopted neuroendocrine-immune regulation of BIMS to trigger dynamic scheduling and AGV repair operations. When $t =25$ s, AGV2 was prepared to perform transportation task 31. Since AGV2 failed before the workpiece was taken, transportation task 31 was sent back to AS/RS. Then AS/RS re-allocated transportation task 31 to AGV1 by using a scheduling approach based on hormone regulation mechanism. During the recovery period of AGV2 (25–60 s), only AGV1 performed transportation operation. When AGV2 recovers from failure, BIMS returns to the normal production.

As shown in Fig. 5.9, at $t =32$ s, BIMS receives a set of rush orders (order 3). By using immune monitoring function BIMC in shop floor level quickly detected the disturbance, and issued a task stop notification. At this time, AGV1, M1, M2, M3, and M4 were performing each current task, and order 3 was executed immediately after the current task finished. BIMS used the dynamic scheduling approach inspired by neuroendocrine regulation mechanism to distribute operation tasks. Simultaneously, BIMS adopted on-line scheduling approach based on hormone regulation to assign corresponding transportation tasks. At $t =146$ s the last transportation task of the rush order was released, and AGVs can accept transportation tasks of the non-emergency order. At $t = 160$ s, the rush order is completed and BIMS returns to normal production.

It can be seen from the above experiments that the BIMS disturbance handling approach based on neuroendocrine-immune regulation mechanism proposed in this chapter can quickly and effectively deal with sudden disturbances in a dynamic environment.

5.5.3 Performance Indicator Analysis

In order to verify the superiority of the BIMS disturbance handling approach, the proposed approach is compared with the dynamic scheduling approach based on MAS. In this section, the same disturbances are introduced to both approaches and in the same experiments described below:

(1) In the MAS approach, under normal circumstances, the approach utilizes negotiation and bidding mechanisms between agents to solve the on-line distribution problem of operation tasks and transportation tasks in the manufacturing system. When disturbances occur, the MAS regards disturbance tasks as new task agents to assign through communication, bidding and negotiation between machine and AGV agents.

(2) In the BIMS approach, under normal conditions, BIMCs at different levels are organized into hierarchical structures, and use the neuroendocrine regulation mechanism and hormone regulation mechanism to distribute operation tasks and transportation tasks on-line. When disturbances occur, BIMCs uses immune monitoring function to quickly detect disturbances, and generate strategies through immune response function. Then the treatment of disturbances can be handled according to neuroendocrine-immune mechanism.

Different orders arrive at manufacturing systems in sequence, and various tasks belonging to the same order arrive at manufacturing systems at the same time. The simulation programs of the two approaches were run 25 times, respectively, and then the experimental data were extracted. Quantitative performance indicators can be found in Fig. 5.10 and qualitative performance indicators can be found in Fig. 5.11. From the comparison of quantitative performance indicators, it can be seen that the two approaches are very close to each other under normal conditions; in case of disturbances, resource utilization rate and output rate of BIMS approach are higher than MAS approach, and cycle time of BIMS approach is lower than MAS approach. Therefore, the qualitative performance indicators of the BIMS approach are better than MAS approach under the disturbance condition, which embodies better production optimization ability. It can be seen from the comparison of qualitative performance indicators that the BIMS approach has a lower loss of output rate than the MAS approach, thus showing higher agility.

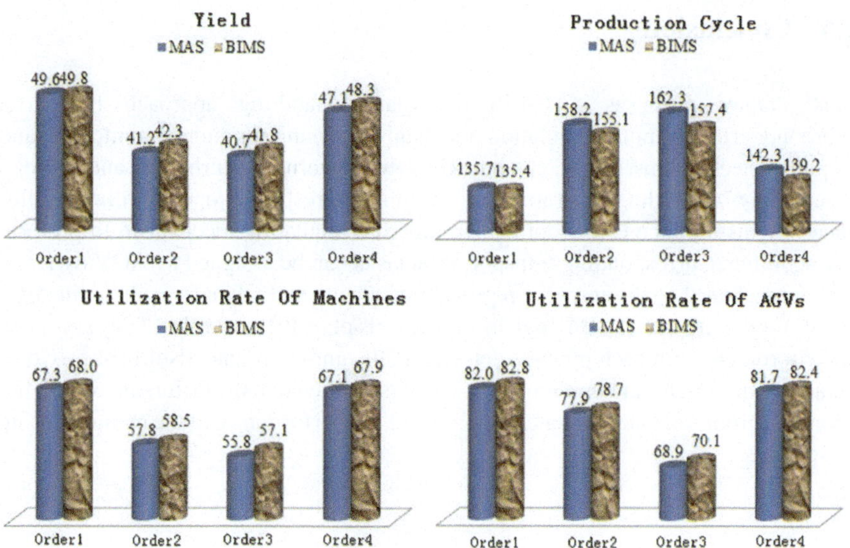

Fig. 5.10 Comparison of quantitative performance indicators

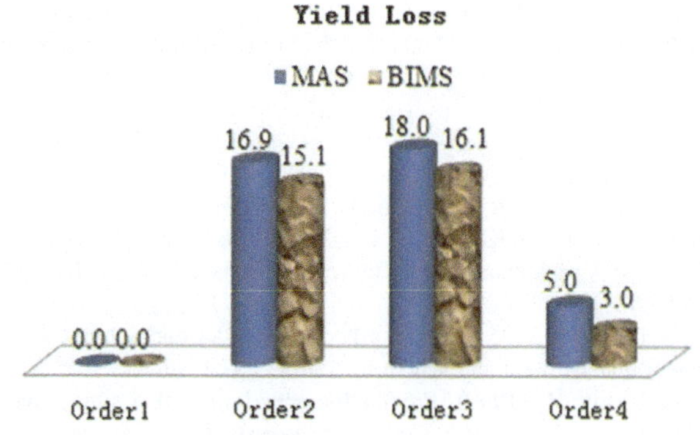

Fig. 5.11 Comparison of qualitative performance indicators

By analyzing quantitative and qualitative performance indicators, the BIMS approach shows better performance than the MAS approach in dealing with disturbances. The results of comparative experiments show that the BIMS disturbance handling approach based on neuroendocrine-immune proposed in this chapter has a good potential to improve the performance of the manufacturing system.

5.6 Conclusion

This chapter proposes a BIMS disturbance handling approach based on neuroendocrine-immune regulation mechanism. Using immune monitoring and response mechanisms, BIMC can quickly detect external disturbances and respond accordingly. According to strategies from immune response, operation tasks influenced by disturbance can adopt neuroendocrine regulation mechanism to preform rescheduling. Corresponding transportation tasks can be assigned by AGV dynamic scheduling based on the hormone regulation mechanism. In dynamic scheduling processes for disturbances, BIMCs at different levels play different roles. They stimulate and coordinate with each other to detect, handle, and eliminate disturbances. Comparative experiments are implemented, and it is verified that the disturbance handling approach proposed in this chapter can improve the performance of the manufacturing system.

References

1. Cowling, P., & Johansson, M. (2002). Using real time information for effective dynamic scheduling. *European Journal of Operational Research, 139*(2), 230–244.
2. Suresh, V., & Chaudhuri, D. (1993). Dynamic scheduling—A survey of research. *International Journal of Production Economics, 32*(1), 53–63.
3. Stoop, P. P. M., & Wiers, V. C. S. (1996). The complexity of scheduling in practice. *International Journal of Operations & Production Management, 16*(10), 37–37.
4. Vieira, G. E., Herrmann, J. W., & Lin, E. (2003). Rescheduling manufacturing systems: A framework of strategies, policies, and methods. *Journal of Scheduling, 6*(1), 39–62.
5. Aytug, H., Lawley, M. A., McKay, K., et al. (2005). Executing production schedules in the face of uncertainties: A review and some future directions. *European Journal of Operational Research, 161*(1), 86–110.
6. Herroelen, W., & Leus, R. (2005). Project scheduling under uncertainty: Survey and research potentials. Project management and scheduling, 289–306. Elsevier.
7. Mehta, S. V., & Uzsoy, R. (1999). Predictable scheduling of a single machine subject to breakdowns. *International Journal of Computer Integrated Manufacturing, 12*(1), 15–38.
8. Yang, Hongbin, & Yan, Hongsen. (2010). Deadlock-free scheduling of knowledgeable manufacturing cell with limited buffers. *Systems Engineering Theory and Practice, 30*(12), 2259–5568.
9. Lei, J., & Yang, Z. Y. (2013). Disturbance management design for a holonic multiagent manufacturing system by using hybrid approach. *Applied Intelligence, 38*(3), 267–278.
10. Merdan, M., Moser, T., Sunindyo, W., et al. (2013). Workflow scheduling using multi-agent systems in a dynamically changing environment. *Journal of Simulation, 7*(3), 144–158.
11. Lou, P., Liu, Q., Zhou, Z., et al. (2012). Multi-agent-based proactive-reactive scheduling for a job shop. *The International Journal of Advanced Manufacturing Technology, 59*(1–4), 311–324.
12. Leitão, P. (2011). A holonic disturbance management architecture for flexible manufacturing systems. *International Journal of Production Research, 49*(5), 1269–1284.
13. Jana, T., Naskar, S., Paul, S., et al. (2015). Handling machine breakdown for dynamic scheduling by a colony of cognitive agents in a holonic manufacturing framework. *Decision Science Letters, 4*(4), 509–524.

14. Umar, U. A., Ariffin, M. K. A., Ismail, N., et al. (2015). Hybrid multiobjective genetic algorithms for integrated dynamic scheduling and routing of jobs and automated-guided vehicle (AGV) in flexible manufacturing systems (FMS) environment. *The International Journal of Advanced Manufacturing Technology, 81*(9–12), 2123–2141.

15. Erol, R., Sahin, C., Baykasoglu, A., et al. (2012). A multi-agent based approach to dynamic scheduling of machines and automated guided vehicles in manufacturing systems. *Applied Soft Computing, 12*(6), 1720–1732.

16. Zheng, K., Tang, D. B., Giret, A., et al. (2015). Dynamic shop floor re-scheduling approach inspired by a neuroendocrine regulation mechanism. *Proceedings of the Institution of Mechanical Engineers, Part B: Journal of Engineering Manufacture, 229*(S1), 121–134.

17. Zheng, K., Tang, D. B., Giret, A., et al. (2018). A hormone regulation–based approach for distributed and on-line scheduling of machines and automated guided vehicles. *Proceedings of the Institution of Mechanical Engineers, Part B: Journal of Engineering Manufacture, 232*(1), 99–113.

18. Cavalieri, S., Macchi, M., & Valckenaers, P. (2003). Benchmarking the performance of manufacturing control systems: Design principles for a web-based simulated test bed. *Journal of Intelligent Manufacturing, 14*(1), 43–58.

Chapter 6
Development of Simulation Platform for BIMS

6.1 Introduction and Synopsis

Dynamic scheduling approaches of the manufacturing system under uncertain disturbance are the main research topics of this book. From Job shop dynamic scheduling approach, on-line AGV scheduling approach to disturbance handling approach of manufacturing systems, each approach is implemented under the same manufacturing system model—bio-inspired manufacturing system (BIMS). Similar to current advanced intelligent manufacturing models (such as MAMS, HMS, FrMS, BMS, etc.), BIMS also adopts a distributed manufacturing system architecture, composed of intelligent units with decision-making and self-governing capabilities, and can achieve overall goals according to coordination-based specific rules between intelligent units. The distributed architecture and autonomy and regulation characteristics of intelligent manufacturing systems empower their inherent advantages in dealing with dynamic events in manufacturing systems. It is also a research hotspot in the field of intelligent manufacturing systems. At present, researches in intelligent manufacturing systems focus on the design of concepts and models, and most of them adopt software simulation in system verification. Although software simulation has the advantages of fast implementation and low cost, it is difficult to reflect a real and dynamic environment of the manufacturing system.

In order to verify the rationality and effectiveness of BIMS approaches and the characteristics of the neuroendocrine-immune mechanism between BIMCs, based on previous research [1–4] our team uses a variety of technologies to design and build a BIMS simulation platform that can simulate the real-time dynamic operating environment of the manufacturing system.

© Springer Nature Singapore Pte Ltd. 2020
D. Tang et al., *Adaptive Control of Bio-Inspired Manufacturing Systems*,
Research on Intelligent Manufacturing,
https://doi.org/10.1007/978-981-15-3445-4_6

6.2 Simulation Platform Architecture

The BIMS simulation platform consists of two sub-platforms: a physical simulation platform and a software simulation platform. In order to combine two sub-platforms, a BIMS simulation platform architecture is designed, as shown in Fig. 6.1. The platform architecture is composed of four layers which are interaction layer, software simulation platform layer, communication layer, and physical simulation platform layer. The interaction layer provides interfaces to operators; the software simulation platform layer provides software operation and simulation functions for BIMS, and realizes operations for server and database; the communication layer establishes

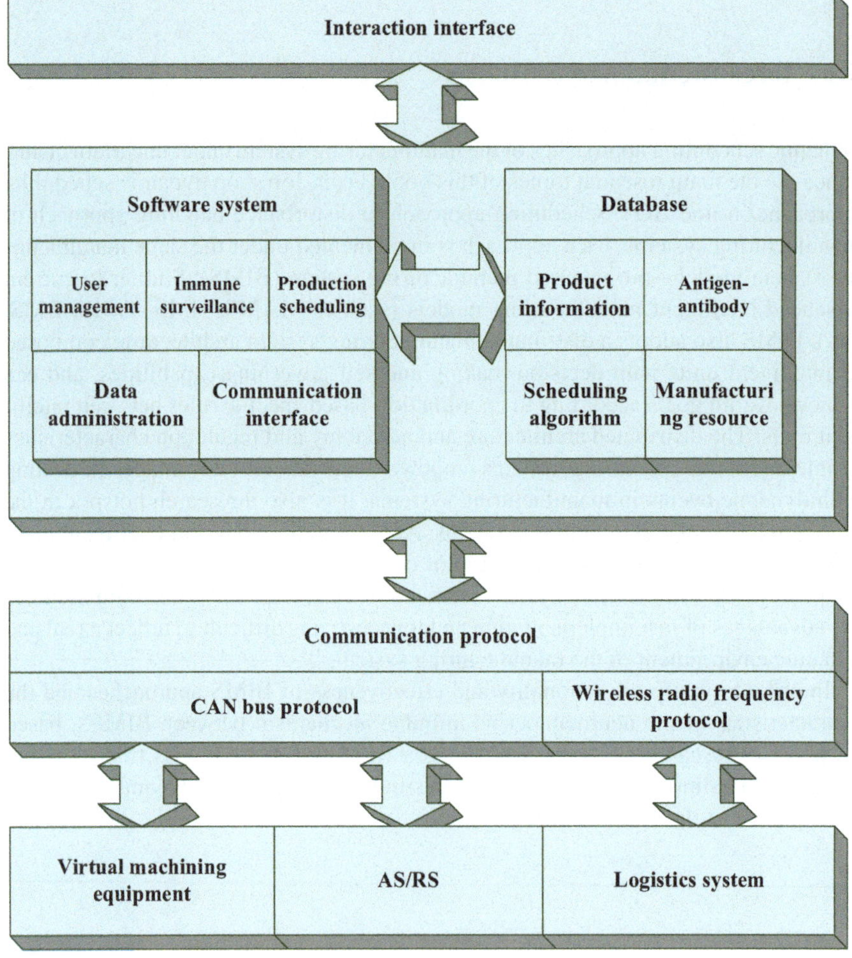

Fig. 6.1 The BIMS system simulation platform architecture

communication protocols between various BIMCs and implements information interaction; the physical simulation platform layer realizes physical entity simulation of each BIMC.

Physical entities of BIMCs (virtual machine, AS/RS, and AGVs) adopt embedded hardware system based on the distributed architecture of BIMS, and perform information interactions to complete their tasks by communication protocol (CAN bus protocol and wireless radio frequency protocol). The software simulation platform utilizes the human–computer interaction interface to monitor the physical simulation platform in real time, and can change the operation strategy of the physical simulation platform through the operator configuration system. At the same time, our team develops software modules according to standards of the physical simulation platform and integrates them into the BIMS simulation platform in the form of Plug and play (PP).

6.3 Physical Simulation Platform

6.3.1 Physical Simulation Platform Architecture

Comparing the manufacturing system with human body system, it can be found that there are many similarities in architecture and function realization. In order to verify the superiority and feasibility of the BIMS architecture, regulation method and scheduling technology are proposed in this book. The architecture of BIMS physical simulation platform (Fig. 6.2) is designed, and radio-frequency communication technology, embedded control technology, CAN bus technology (Controller

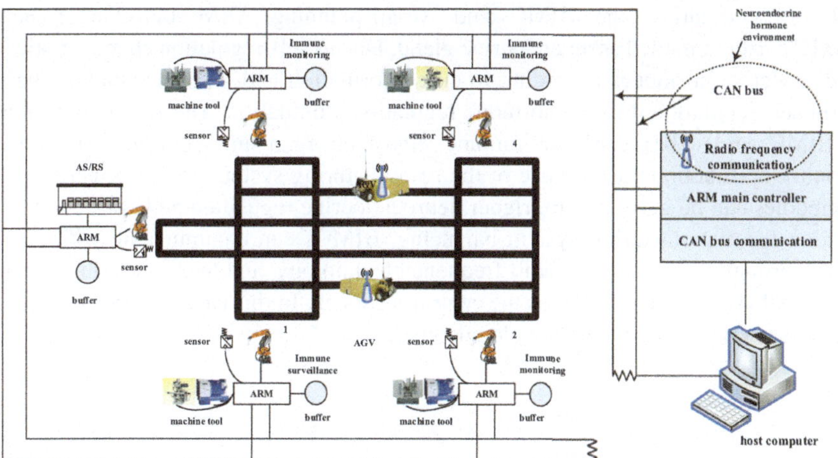

Fig. 6.2 The architecture of BIMS physical simulation platform

Area Network), road identification technology, etc., are integrated into platform development.

BIMS is an intelligent manufacturing system model composed of distributed autonomous BIMCs. The physical entities of BIMC are composed of the following modules: machine tool, AS/RS, automatic guided vehicle (AGV), and master controller. All physical modules are designed and developed by micro controller unit (MCU) based on ARM (Advanced RISC Machine) architecture.

The functions of each physical module are as follows:

(1) Master control module consists of a host computer and an ARM master controller. The host computer realizes human–computer interaction through software, and obtains data from the physical simulation platform through the communication interface. The ARM master controller realizes the control and coordination between physical modules through the multi-task processing function.

(2) Machine tool module consists of an ARM controller, sensor, virtual equipment, robot, and buffer, which realizes the simulation of machine tool operating.

(3) AS/RS module is composed of ARM controller, sensor, robot, and buffer, which performs automatic reclaiming and storage of raw materials and finished products.

(4) Logistics module is composed of AGV equipped with ARM controller and performs material handling through radio frequency communication.

A BIMC in physical simulation platform is composed of decision-makers, sensors and controllers, imitating regulation of the neuroendocrine-immune systems. Self-regulation and control are carried out within the external environment change, to realize adaptive response to a complex manufacturing system in a dynamic environment. Similar to the human body system, the ARM master controller in a physical simulation platform can be regarded as the central nervous system, which plays the role of coordinating other BIMCs and overall planning. ARM controller of each BIMC can be regarded as an endocrine gland. Due to self-regulation characteristics, it can secrete hormones according to task information and its own state, and perform self-regulation through hormone regulation stimulation. The ARM controller of BIMC can also be regarded as immune cells of the organism. According to immune monitoring function, disturbance of the manufacturing system can be detected and antibodies can be generated to trigger neuroendocrine regulation and hormone regulation for manufacturing system balancing. BIMCs can communicate with each other through CAN bus and radio-frequency technology, and feedback information to the ARM master controller as the central feedback. In this case, BIMS can adjust and control production from the global perspective.

6.3.2 Quasi-hormone Communication Protocols

In the physical simulation platform, ARM controllers of BIMCs communicate with each other according to corresponding communication protocols. A quasi-hormone wireless communication protocol is used to communicate between the master controller and AGVs. The CAN-based quasi-hormone communication protocol is used to implement communication between master controller, machine tools, and the warehouse. Two communication protocols are described in detail below.

6.3.2.1 Quasi-hormone Radio Frequency Communication Between Master Controller and AGVs

In this chapter, quasi-hormone communication protocols are written into embedded programs. The master controller and ARM in AGV communicate through the radio-frequency communication module, and use Serial Peripheral Interface (SPI) to send and receive signals. Serial port signals of the master controller are transformed into wireless signals by radio frequency module and sent to AGV in the form of quasi-hormone. Similarly, AGV wireless signal is transformed into serial port signal through radio-frequency module and sent to the master controller. The standard serial communication protocol of quasi-hormone communication is used to process data of radio-frequency communication module. Information format is shown in Table 6.1.

Since information is sent in both directions and the sender and receiver are not unique, it is necessary to specify the data format of the information. This communication protocol specifies the sender address, receiver address, command field, and data field in a message, and each segment of information use 0xAA and 0x55 as the frame header. The master controller and AGV are receivers to each other, where the AGV address is 0x07 or 0x08 and the master controller address is 0x06. In order to identify the direction of information transmission, the command field is set in the communication protocol. The command field 0x01 represents information being transmitted by the master controller to AGV, and 0x02 represents information being propagated by AGV to the master controller. Data fields represent specific contents for sending. When the command field is 0x01, data field includes two bytes, representing AGV's destination (01: to machine 1, 02: to machine 2, 03: to machine 3, 4: to machine 4, 05: to warehouse) and the number of parts carried by AGV. When the command field is 0x02, data field also contains two bytes, representing coordinates of AGV X and Y. The control system verifies the correctness of the information by setting "checksum", which calculates the XOR value of all bytes from the beginning

Table 6.1 Information format of quasi-hormone communication

Frame header	Frame length	Receiver address	Sender address	Command field	Data field	Checksum
2 bytes	1 byte	1 byte	1 byte	1 byte	N bytes	1 byte

Table 6.2 Quasi-hormone communication protocol between the master controller and AGV

0x55 0xAA	Frame header, represent command arrival
0x05	Frame length, followed by 4 character commands
0x07 etc.	AGV address
0x06	Main controller address
0x01	Command: ARM notifies AGV to go to appropriate locations
0x02	Command: AGV reports current coordinates to ARM
0x01 etc.	Target address 01: to machine 1; 02: to machine 2; 05: to warehouse
0x03 etc.	Workpiece number
0x02 etc.	AGV current position: X coordinate
0x04 etc.	AGV current position: Y coordinate
0xf8 etc.	XOR value of all bytes from the beginning of a frame to the end of data field

of the frame to the end of the data field. The quasi-hormone communication protocol between the master controller and AGV is shown in Table 6.2. The hormone information fragments represent functions in the protocol.

In the process of radio-frequency communication, AGV interprets information under hormone regulation of the master controller, and goes to destinations according to commands of the master controller through road identification. AGVs can also record their own coordinates and feedback to the master controller through quasi-hormone communication. After the system is powered on, the program will be initialized, and AGVs will return to origin positions. When AGVs receive transportation orders according to destination addresses, relaying on the infrared sensor tracking, AGV decorated along the map route to specified locations. Simultaneously, AGVs will feedback their own information to the master controller in real time.

6.3.2.2 CAN Communication Between the Master Controller and Devices

In the physical simulation platform, CAN bus (hormone) is used to communicate between the master controller (central nervous system), warehouse and machine tool controllers (endocrine gland). Although machine tools and warehouse controllers have certain local autonomy ability, the master controller still needs to monitor the real-time state of each device, so as to carry out global level adjustment. To this aim, this chapter designs a quasi-hormone communication protocol based on CAN bus. The total length of the message is seven bytes, and each bit represents corresponding hormone information. By decoding each bit of message frame one by one, information content transmitted between controllers can be obtained, shown in Table 6.3.

Table 6.3 Quasi-hormone communication protocol based on CAN bus

Bit0–Bit3	Workpiece number (supports 16 types)
Bit4–Bit7	Robot number (supports 16 types)
Bit8–Bit15	Operation time of the process
Bit16	0: automatic mode; 1: manual mode
Bit17	0: normal warehouse delivering; 1: special warehouse delivering
Bit18	0: non-hormone data transmission; 1: hormone data transmission
Bit19	Robot accepts information (1) or sends information (0)
Bit20	AGV arrived at the correct destination: 1: arrived; 0: did not arrive
Bit21	Device state: 1: fault; 0: normal
Bit22	Status information sending bit: 1: sent information is device status information; 0: non-device status information
Bit23	Robot grabs a workpiece (1); robot unloads workpiece (0)
Bit24–Bit31	Hormone information caused by operation tasks
Bit32–Bit39	Hormone information caused by transportation tasks
Bit40–Bit47	AGV hormone information
Bit48–Bit55	Other uses

The communication protocol can achieve the following functions: when an external task (stimulus) enters the manufacturing system, the master controller (central nervous system) immediately responds to the task (stimulus) and issues command information (quasi-hormone) according to quasi-hormone communication protocol based on CAN bus. The warehouse (gland) receives information and performs operations (delivering or storing workpiece); when the warehouse receives an AGV positioning information (quasi-hormone), it will control the robot to load or unload parts of AGV, and feedback status information (quasi-hormone) to the master controller, which uses quasi-hormone radio frequency communication for completing corresponding transportation tasks. Machine tools also communicate with the master controller through the quasi-hormone communication protocol based on CAN bus, and analyzes communication contents to perform corresponding autonomous behaviors. In this way, the basic operations of loading, processing, and unloading in machine tools can be realized. BIMS adopts these two hormone communication protocols to establish a communication network between controllers (ARM), which can complete the autonomy and regulation of BIMS.

6.3.3 Physical Simulation Platform

According to the concept of the physical simulation platform and quasi-hormone communication protocols, our team built a BIMS physical simulation platform

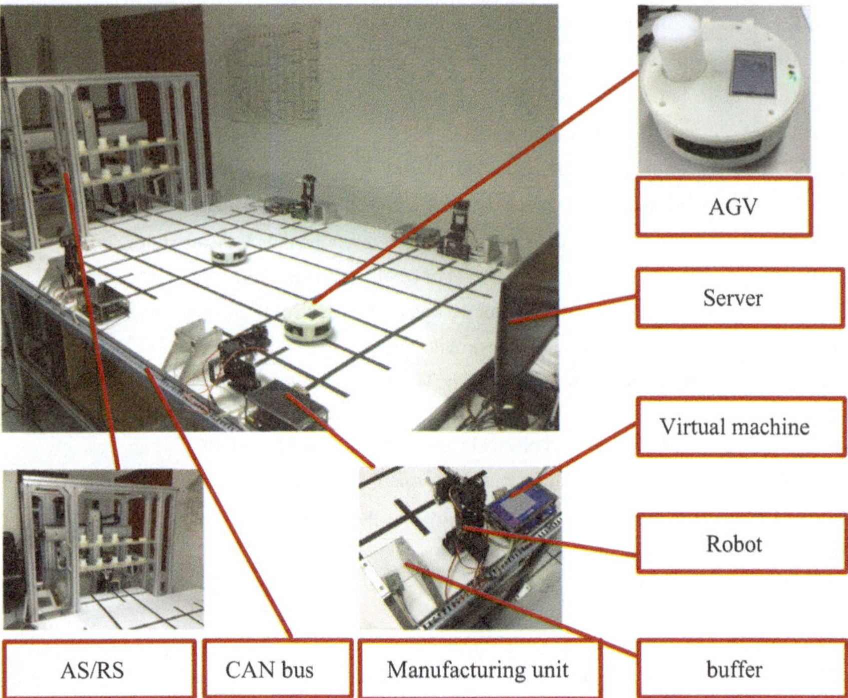

Fig. 6.3 The BIMS physical simulation platform

by using various technologies (including embedded control technology, radio-frequency communication technology, CAN bus communication technology, and road identification technology), shown in Fig. 6.3.

From the perspective of neuroendocrine-immune regulation, the ARM master controller in the physical simulation platform is similar to the central nervous system. ARM controllers of each BIMC are similar to endocrine glands (cells); CAN bus and radio-frequency communication protocols constitute the quasi-hormone regulatory environment of the physical simulation platform. Meanwhile, the ARM master controller and other ARM controllers have similar functions of immune cells. The communication protocols and regulation strategies between ARM controllers formulate the neuroendocrine-immune regulation rules of BIMS.

In summary, compared with other modern intelligent manufacturing systems, the BIMS physical simulation platform built in this chapter has the following advantages and significance:

(1) This simulation platform is not only used for computer simulation but can also reflect the production process in manufacturing system more truly.

(2) On the physical simulation platform, executions of different regulation mechanisms, autonomy of BIMCs and self-organization behaviors between BIMCs,

and self-organization and self-adaptation of BIMS in the disturbed environment can be verified.

(3) The physical simulation platform uses ARM as a controller which has characteristics of small size, low power consumption, modularity convenient for subsequent development, and improvement.

6.4 Software Simulation Platform

In order to apply dynamic scheduling approaches to the physical simulation platform, our team designs and develops a BIMS software simulation platform, which takes Visual Studio 2010 of Microsoft as the development environment, uses C# for programming and Microsoft Access for the database.

6.4.1 Software Simulation Platform Architecture

An overall architecture of the software simulation platform is shown in Fig. 6.4. It consists of user management module, data management module, immune monitoring module, production scheduling module, and communication interface module.

The content of each functional module is described as follows:

(1) User management module: it completes functions of user login, permission management, and password management, and ensures the stability and security of software system by managing and controlling permissions of different operators.

(2) Data management module: it completes data information management in the manufacturing system, including task information management module, antigen and antibody management module, scheduling algorithm management module. Task information management module manages system access for task information. Antigen and antibody management module manages disturbance index of the manufacturing system and antigen and antibody rules. Scheduling algorithm management module can view the existing scheduling algorithms embedded in the physical simulation platform and the server, and add/remove existing scheduling algorithms through the software simulation platform.

(3) Immune monitoring module: it can display operation status and data of each BIMC in real time for administrators viewing. It can recognize disturbances according to immune response mechanisms, generate antigens and antibodies, and provide a basis for agile response to disturbances.

(4) Production scheduling module: scheduling algorithm configuration module can realize the selection and configuration of scheduling algorithms used by the manufacturing system in different situations.

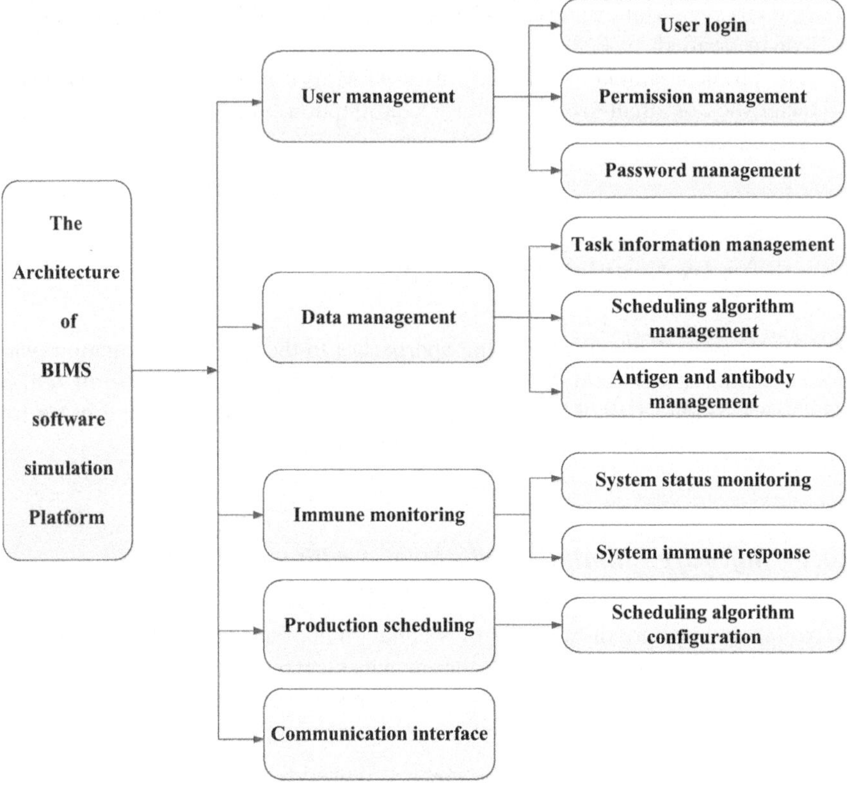

Fig. 6.4 An overall architecture of software simulation platform

(5) Communication interface module: it serves as a bridge between software simulation platform and physical simulation platform, which is used to transfer data and information between software simulation platform and scheduling software, and collect real-time information form physical entities.

6.4.2 Function Modules of Software Simulation Platform

The main interface of the software simulation platform is shown in Fig. 6.5, including operation management, information management, data management, system parameter setting, dynamic events, production scheduling, monitoring, and communication functions.

The main interface displays the basic information of the BIMS simulation platform including 4 machine tool BIMCs, 2 AGV BIMCs, and 1 warehouse BIMC. Each BIMC displays its own operating and monitoring status. Click the loading system

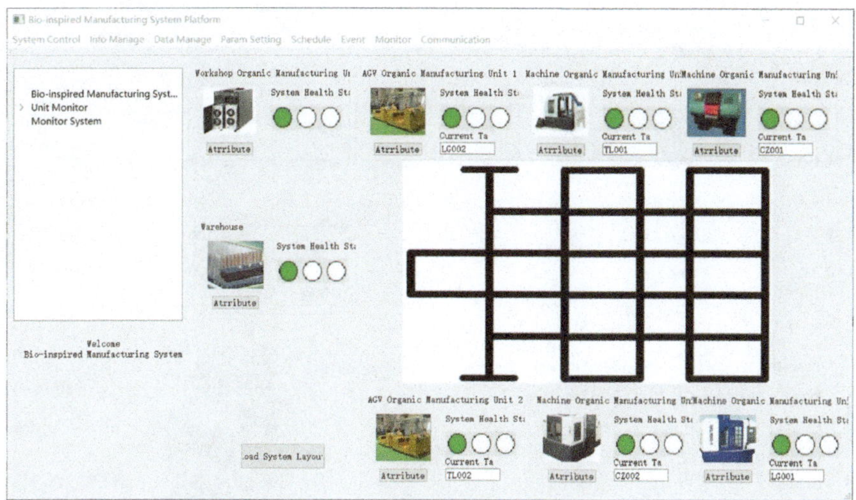

Fig. 6.5 A main interface of the software simulation platform

layout button to load the path layout between machine tools and AGVs, and system layout can be changed according to different experimental requirements. Data management module, immune monitoring module, and production scheduling module of the software simulation platform will be described in detail in the following subsections.

6.4.2.1 Data Management Module

The data management module includes task information management, antigen and antibody management, scheduling algorithm management, and other functions.

With continuous starting and end of production tasks in BIMS, task information changes correspondingly. All task information can be displayed in the task list as shown in Fig. 6.6. Operators can edit and modify task information saved in the database.

In BIMS, due to randomness of uncertain disturbance events, the immune monitoring module determines the type of disturbance through antigen recognition and generates corresponding antibodies. For antigens and antibodies produced by unknown disturbances, they need to be input into the antigen/antibody library. Using this approach, BIMS can react quickly when the same disturbance occurs next time.

The antigen and antibody management module is shown in Fig. 6.7. The operator can add new antigen/antibody data into the library which can be managed by viewing, modifying, and deleting operations.

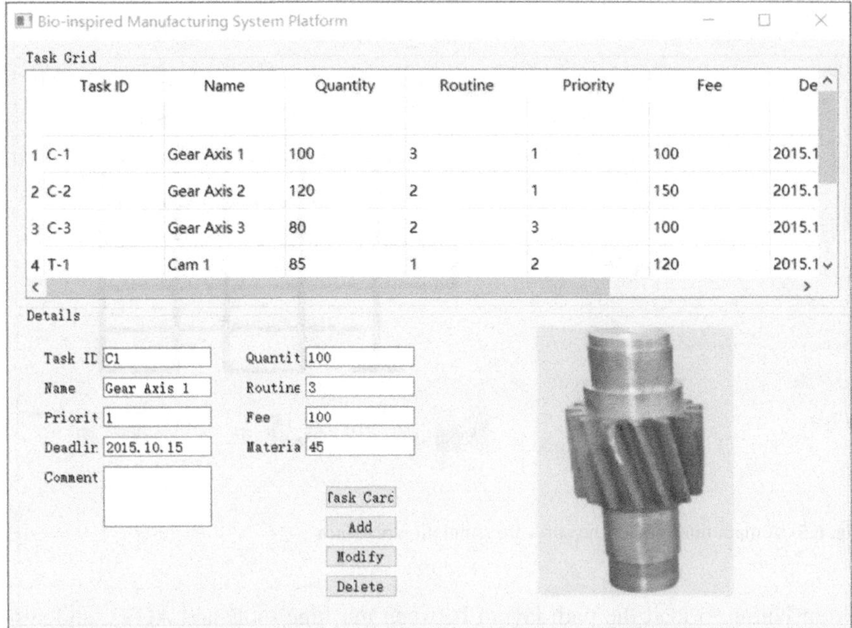

Fig. 6.6 Task information management module

During the research on BIMS, our team accumulated several scheduling algorithms [4–9]. Each algorithm has its own characteristics and application range. When performing scheduling comparison experiment, comparison algorithms can be added to scheduling algorithm library to facilitate switching of different scheduling algorithms for verification as shown in Fig. 6.8. Operators can view specific information on different scheduling algorithms and set up scheduling algorithms.

6.4.2.2 Immune Monitoring Module

The Immune monitoring module of BIMS completes disturbance recognition, antigen and antibody matching, antigen concentration calculation and immunity index calculation when a disturbance occurs as shown in Fig. 6.9. Equipment status, number of current processing task, and operation information affected by the next processing tasks can be viewed and checked. And the occurrence of system disturbance events, antigen and antigen concentration, antigen and antibody status, and system immunity index can also be checked.

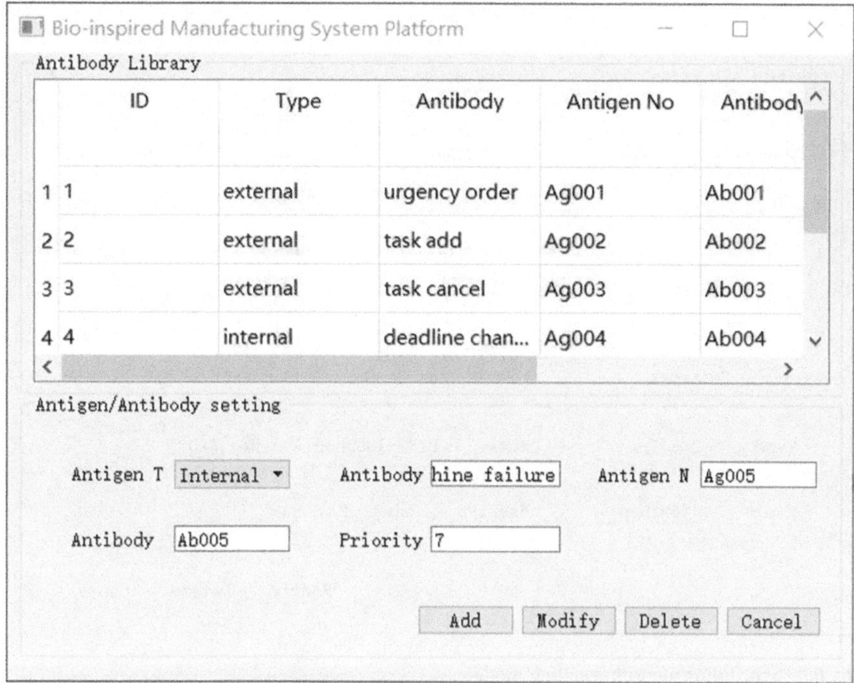

Fig. 6.7 Antigen and antibody management module

6.4.2.3 Production Scheduling Module

The production scheduling module is to configure different scheduling algorithms according to different requirements. As shown in Fig. 6.10, scheduling algorithms for all BIMCs can be viewed, and required scheduling algorithms can be configured. When BIMS needs a highly optimized scheduling result and no requirement for computing time, an off-line scheduling algorithm with strong global optimization ability is usually selected for the BIMC in the shop floor level. In the case of system disturbance, real-time and fast scheduling is required. The operator can allocate on-line scheduling algorithms according to the requirements of antibody to eliminate the disturbance agility.

In the structure tree of the main interface, operation status of each BIMC in BIMS can be viewed and it is shown in Fig. 6.11. During simulation system operation, the status information of each BIMC can be collected through CAN bus and displayed, such as equipment type, equipment status, starting time, processing time, equipment utilization rate, etc. When equipment fails or is repaired, corresponding time node will also be displayed. When a BIMC participates in dynamic scheduling based on hormone regulation, hormone information of success or failure task will be displayed in the hormone information interface.

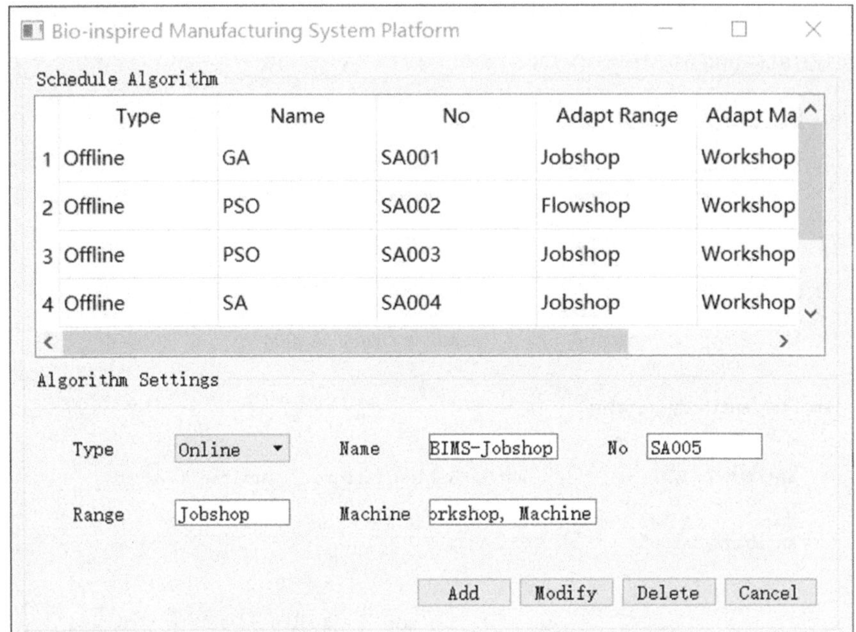

Fig. 6.8 Scheduling algorithm setting module

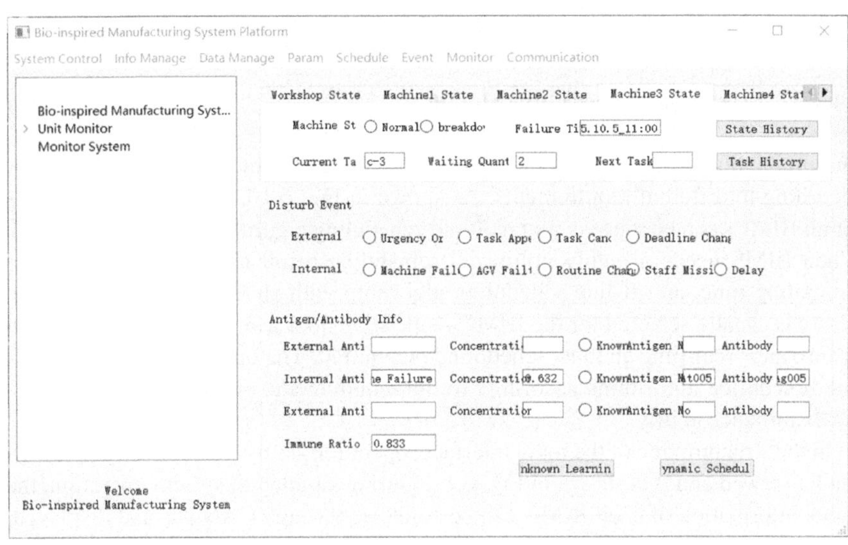

Fig. 6.9 Immune monitoring module

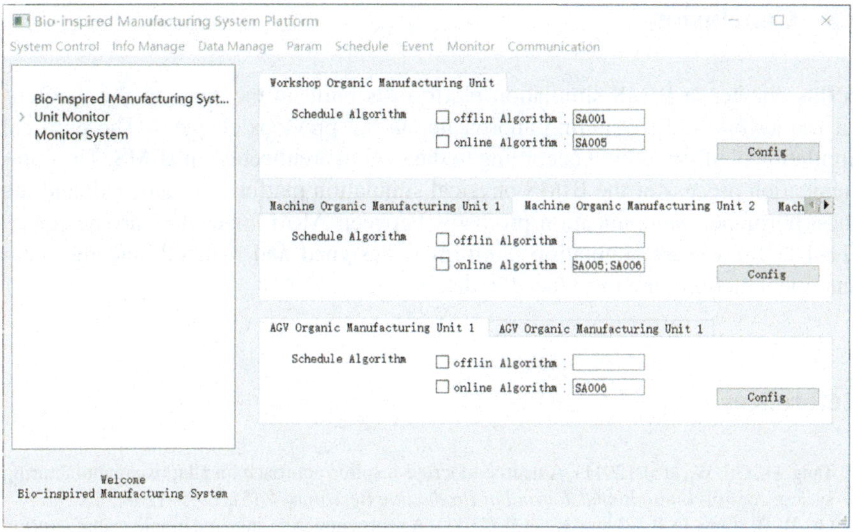

Fig. 6.10   Configuration interface of scheduling algorithms

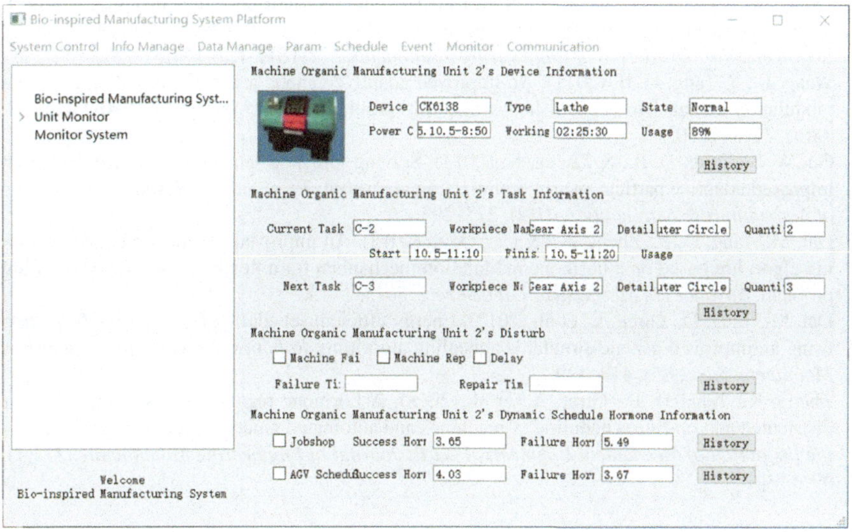

Fig. 6.11   Operation status of BIMC

During the operation of BIMS, the simulation software records information displayed in the interfaces of each BIMC. Operators can view the historical records of information and use the data as the basis for statistics and analysis.

6.5 Conclusion

In this chapter, a BIMS simulation platform is built on the basis of the theoretical framework and scheduling approaches in the previous chapters. The physical simulation platform is built according to the system architecture of BIMS. The communication process of the BIMS physical simulation platform is analyzed, and the quasi-hormone communication protocols between ARM controllers are designed. The BIMS software simulation platform is designed and realized, and the main functional modules are introduced in detail.

References

1. Tang, D., Gu, W., et al. (2011). A neuroendocrine-inspired approach for adaptive manufacturing system control. *International Journal of Production Research, 49*(5), 1255–1268.
2. Gu, W. B., Tang, D. B., Zheng, K., et al. (2011). A neuroendocrine-inspired bionic manufacturing system. *Journal of Systems Science and Systems Engineering, 20*(3), 275–293.
3. Tang, D. B., Zheng, K., Zhang, H. T., et al. (2018). Using autonomous intelligence to build a smart shop floor. *The International Journal of Advanced Manufacturing Technology, 94,* 1597–1606.
4. Zheng, K., Tang, D. B., Giret, A., et al. (2015). Dynamic shop floor re-scheduling approach inspired by a neuroendocrine regulation mechanism. *Proceedings of the Institution of Mechanical Engineers, Part B: Journal of Engineering Manufacture, 229*(S1), 121–134.
5. Wang, L., & Tang, D. B. (2011). An improved adaptive genetic algorithm based on hormone modulation mechanism for job-shop scheduling problem. *Expert Systems with Applications, 38*(6), 7243–7250.
6. Gu, W. B., Tang, D. B., & Zheng, K. (2014). Solving job-shop scheduling problem based on improved adaptive particle swarm optimization algorithm. *Transactions of Nanjing University of Aeronautics & Astronautics, 31*(2), 275–293.
7. Dai, M., Tang, D. B., Zheng, K., & Cai, Q. X. (2013). An improved genetic-simulated annealing algorithm based on a hormone modulation mechanism for a flexible flow-shop scheduling problem. *Advances in Mechanical Engineering, 2013,* 1–13.
8. Dai, M., Tang, D., Giret, A., et al. (2013). Energy-efficient scheduling for a flexible flow shop using an improved genetic-simulated annealing algorithm. *Robotics and Computer-Integrated Manufacturing, 29*(5), 418–429.
9. Zheng, K., Tang, D. B., Giret, A., et al. (2018). A hormone regulation–based approach for distributed and on-line scheduling of machines and automated guided vehicles. *Proceedings of the Institution of Mechanical Engineers, Part B: Journal of Engineering Manufacture, 232*(1), 99–113.

The manufacturer's authorised representative in the EU is Springer
Nature Customer Service Centre GmbH, Europaplatz 3, 69115 Heidelberg,
Germany. If you have any concerns regarding our products, please
contact ProductSafety@springernature.com

Printed and bound by CPI Group (UK) Ltd, Croydon, CR0 4YY
28/11/2025
02007724-0001